Iain Stewart

The challenges of climate change and its impact on the environment and us are unprecedented. We will only overcome them if our approach is integrated and holistic. Science has to be joined up, and more, scientists and society have to understand each other. And that means that we scientists have to communicate a lot more effectively with society than we do.

This book about the rocks of one small part of our planet - the beautiful county of Northumberland in the UK - is trying to do exactly that. In the descriptions of 50 special places it reaches out beyond the science of geology, to other aspects of the landscape, to biology, and to our shared history and heritage. The plain words and clear, attractive images will engage readers of all backgrounds.

This book seeks to inspire you to be curious about the ground beneath your feet and explore the landscape for yourself. Turn the pages and then pull those boots on, you won't regret it!

Professor Iain Stewart MBE is a geologist who has presented many BBC documentaries on the history and dynamics of the Earth. His passion is to improve the communication of science. He is currently Jordan-UK El Hassan bin Talal Research Chair in Sustainability.

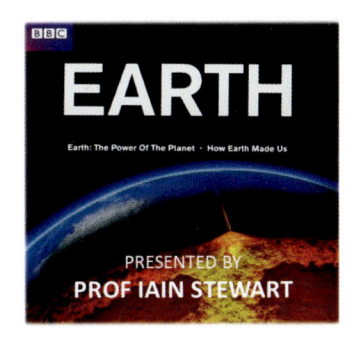

Simon Winchester

I feel passionately that geology is the most romantic of sciences, as well as being of the most profound importance. The days I spent in the field in my early career were for me times of absolute wonder and great happiness. The opportunity to visit beautiful, natural places, in the company of scientists who had the uncanny gift of knowing just why these beautiful places were exactly as they were, remains a great privilege.

*I spent the first three years of my professional life in Northumberland, coming to know the county - Berwick to Alnwick, Morpeth to North Shields, Rothbury to Wallsend - as well as anywhere in England. So I have a particular fondness for **Northumberland Rocks**, which paints a fascinating picture of the evolution of this historic county across geological time. This invaluable book ranges, just as I did fifty years before, from the Scottish border, to the edge of the Roman Empire, to the centre of a great and ancient city, and it profits from a scientist's insight, uncomplicated language and delightful images, to reveal why this magnificent and remarkable landscape is as it is.*

Simon Winchester OBE is a British-American author and geologist. He was a journalist for The Guardian and now the New York Times. His internationally best selling books include The Map that Changed the World and The Professor and the Madman.

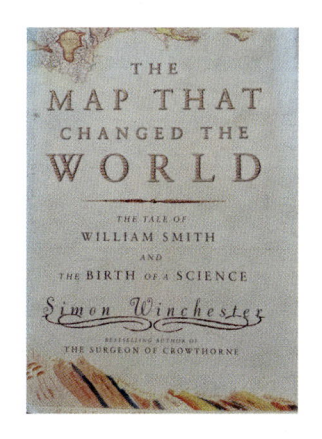

Foreword

Compare a map of natural habitats with a geological map of the same area anywhere on our planet and you will immediately see how strong the link is between geology and wildlife. The rocks beneath our feet are a key factor in the distribution of habitats and species. Geodiversity underpins biodiversity, which explains why Northumberland Wildlife Trust, in common with Wildlife Trusts across Britain, includes conservation, protection and promotion of geology in its founding principles.

Our ambitions to conserve and enhance wildlife depend on a holistic understanding of our landscape – whether that is a Border Mire, a Cheviot mountain, the coast, or an inner-city park. Rocks shape our country and form our soils; they are the natural connector as we establish and extend nature networks. Throughout history rocks have influenced the way we humans have used the land. The impacts of our changing climate and the complex threats to our environment concern us all. If we better understand the diverse factors involved we will be better equipped to meet those challenges and fight for nature's recovery.

This book sets out to help us understand the bedrock of our landscape - literally. It not only celebrates the amazing geodiversity of our region, through photographs and simple descriptions of 50 special places, it makes the story of the Northumbrian landscape relevant and more accessible to everyone.

Mike Pratt, CEO Northumberland Wildlife Trust

Preface

This book is a personal selection of 50 places across the old county of Northumberland. Places where rocks make a special statement about the region's landscape, history, culture and wildlife.

Some of these places are well known, other less so and the journey to discover them has taken me to hidden corners. The rock sites are an attempt to cover the geography of the historic county, but more to explore its third dimension, its geological past. This book aspires to make that story accessible; it is intended for everyone who is curious about their landscape.

What started out as desire to engage friends and non-geologist colleagues in the amazing stories our rocks tell, morphed into a celebration of 50 years of Northumberland Wildlife Trust, and now into this book. Author's royalties from its sale will go to the Trust.

I hope this book engages you.

Ian Jackson, September 2021

Thank you very much for buying the book! The first and second print runs of 2000 copies each have sold out, so this is the third print run! Who knew a book about rocks would prove so popular? But maybe it shouldn't be a surprise, after all the subject is the stunning landscape of Northumberland!

Ian Jackson, September 2023

Looking east at the Whin Sill of Peel Crags from Steel Rigg

Sources and acknowledgements

The descriptions of rocks and landscapes in this book are drawn from many sources, but principal amongst them are the maps, memoirs and reports of the field geologists of the British Geological Survey.

Angus Lunn, geologist, ecologist and author of several books about the natural history of Northumbria, supported the venture throughout. He kindly coordinated the words on the wildlife and reviewed the text.

The Northumbrian geology excursion guides produced by Colin Scrutton were my sources for coastal sections. Google and Wikipedia were used extensively to locate historical and other information.

Any errors in the descriptions in this book are mine.

Thank you to my wife Gill, and to walking friends, for allowing me to hijack our hikes. The photographs were taken with my iPhone. I'm hoping the beauty of the Northumberland landscape will overcome my photographic shortcomings.

I would like to thank Northumberland Wildlife Trust who generously supported publication of this book.

Foreshore at Howick Bay

Introduction

Northumberland* is blessed with outstanding rocks and landscape – they are the equal of the best in Britain. As a county we probably owe more to our rocks than anywhere else. Without the Whin Sill's crags and cliffs, and the centuries of mining and quarrying of coal, lead, limestone and sandstone, our heritage would be entirely different.

We may live on a very small part of the Earth's crust, but it is one that has a long and eventful history, some of it shrouded in mystery. The county we know as Northumberland has travelled across the globe and lain at the bottom of oceans. The ground beneath us has been buried to red-hot depths, broken apart by earthquakes, injected with molten magma and frozen under ice sheets, emerging only 15,000 years ago as the landscape we know today.

Some will argue, not unreasonably, that several sites in this book are hardly "rocks". True, but collectively the 50 places tell a story of the full 425 million years of the history of the Northumberland landscape, one that sets in context the brief narrative of humankind.

The sites are divided into five themes, but the division is often arbitrary and several could fit into more than one theme.

Northumberland in this book is the historic county, so it includes the city of Newcastle upon Tyne and North Tyneside too.

Ceiling at Crindledykes limekiln

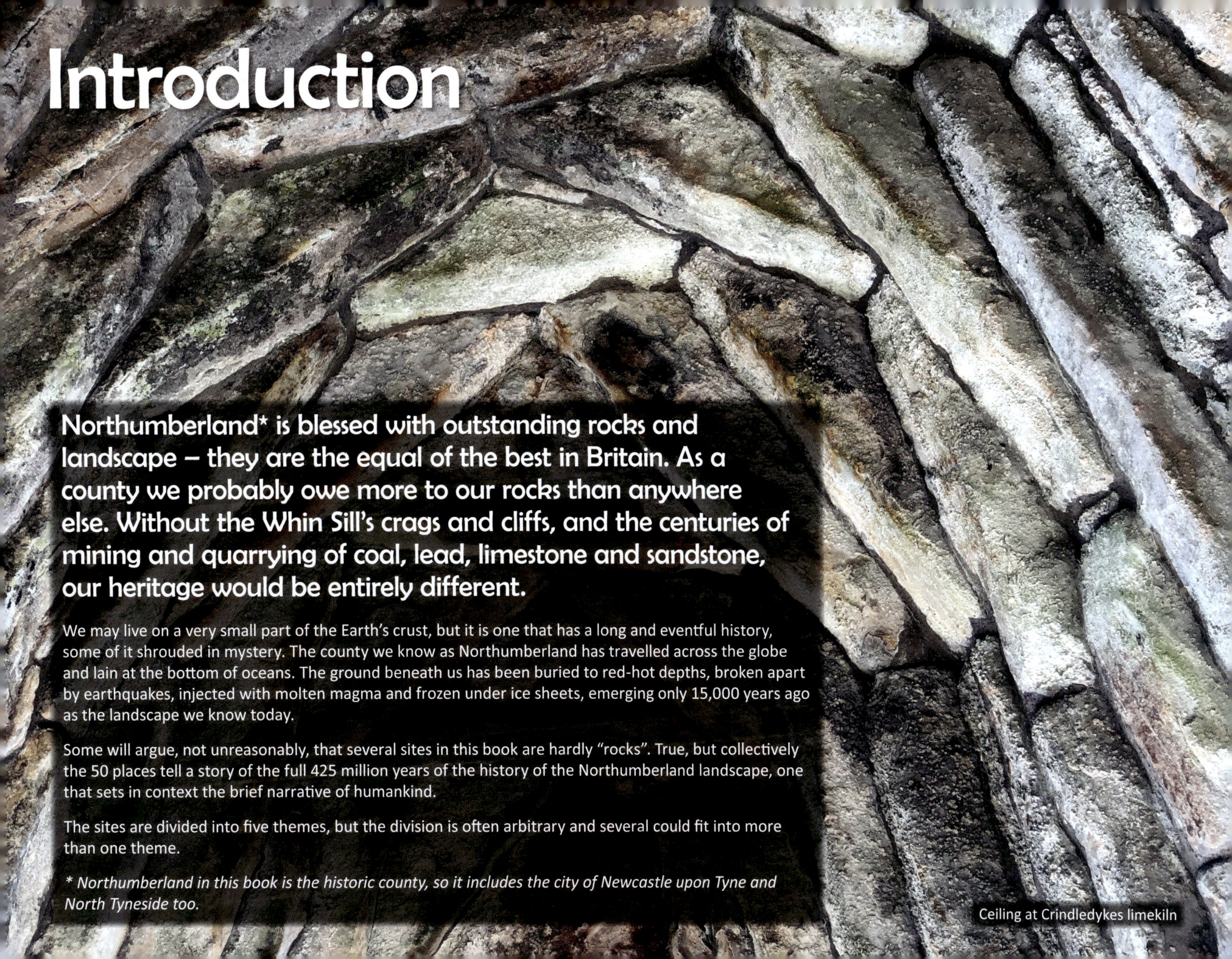

Ancient rivers, seas & life
1 Akenshaw Burn
2 Belsay Hall
3 Drake Stone
4 Green Carts Quarry
5 Jesmond Dene
6 Kielder Stone
7 Queen's Crags
8 Simonside
9 Sugar Sands Bay
10 Tynemouth Castle

Volcanoes & molten rock
11 Coquet valley
12 Cuddy's Crags
13 Cullernose Point
14 Great Standrop
15 Holy Island
16 Housey Crags
17 Limestone Corner
18 Linhope Spout
19 Lumsden Law
20 Walltown Quarry

Earthquakes & folded rocks
21 Berwick
22 Coquet Head
23 Crindledykes
24 Cullercoats Bay
25 Howick Haven
26 Saltpan Rocks

Climate & landscape change
27 Beldon Cleugh
28 Bradford Kame
29 Broomlee Lough
30 Down Hill
31 Druridge Bay
32 Echo Crags
33 Fallowfield
34 Falstone Moss
35 Gelt Linn
36 Hauxley
37 Shaftoe Crags

Heritage & mining
38 Allenheads
39 Beltingham
40 Duddo Stones
41 Dukesfield
42 Hancock Museum
43 Haltwhistle Burn
44 Hareshaw Burn
45 Hartley New Pit
46 Lynemouth
47 Newcastle upon Tyne
48 The Spetchells
49 Town Moor
50 Woodhorn Museum

50 sites in 5 themes

County boundary above Allenheads

Welcome to **NORTHUMBERLAND**
England's Border Country
NORTHUMBERLAND
Northumberland County Council

Northumberland's geology

248
295
Whin Sill
Permian

354
Carboniferous

Cheviot Granite
Devonian

417
Silurian

443

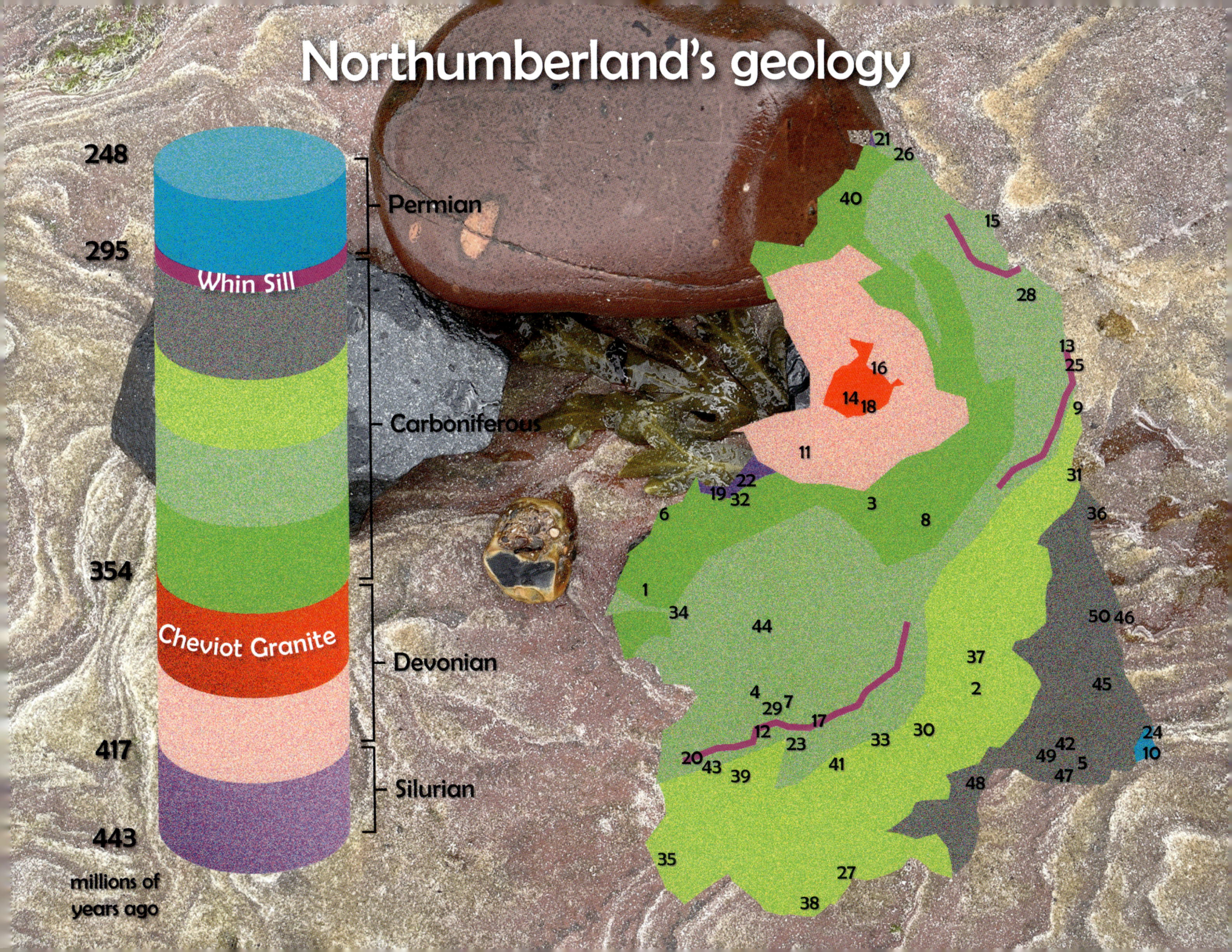

millions of years ago

Akenshaw Burn

Akenshaw Burn runs in a small valley within Kielder Forest Park. Some of the oldest Carboniferous rocks in the county are exposed in the banks of the burn.

The valley is home to willows, hazel and alder but it is within Kielder Forest and surrounding it are huge areas of mainly sitka spruce, grown for timber and planted in the mid-20th century. Before that the valley was part of a very remote hill farm, with moorland of blanket bog and acidic grassland. Roe deer are frequent visitors.

In the burn are bands of two different rocks – grey mudstones and pale brown cementstones. Cementstones are a form of limestone containing magnesium carbonate as well as calcium carbonate. Here the rocks are tilted at angles and broken by a series of geological faults.

350 million years ago the cementstones were lime-rich mud at the bottom of warm salty coastal lagoons that then evaporated. Britain was only 5 degrees north of the Equator then and the climate was much warmer. In many places the thicker cementstone beds have been quarried for local supplies of lime which were used to improve the soil.

A 5 kilometre walk northwest of Matthew's Linn car park on the shore of Kielder Water [NY608896].

Mudstones and cementstones in the bank of the Burn

Belsay Hall and quarry garden

The rock gardens at Belsay Hall were created from an old quarry which was the source of the sandstone used to build the hall and nearby village. The scale of the garden is monumental and harks back to the 19th century fashion for classical romanticism and paying homage to 'awesome nature'.

The sandstones are about 325 million years old from the Carboniferous period. The quarry was active between 1810 and 1817 when the hall was being built. The owner/designer of the hall and garden was Sir Charles Monck. The sandstone is more than 10 metres thick and shows "cross-bedding" really well – that's where the layers in the sandstone are at different angles. That is evidence that it was formed by water, or wind; this sandstone was once sand in an ancient river channel flowing from the north.

The shelter of the high rock walls of the quarry and the damp location provide a perfect site for several plants, many of them alien species. Look for one in particular, called gunnera from Chile. On the rock faces grow broad-buckler fern, hart's-tongue, ivy-leaved toadflax, foxglove, wood-sorrel and wood speedwell. Elsewhere are golden-scaled male-fern, lady fern and pink purslane, with occasional royal fern (once native in Northumberland, this splendid fern has been re-introduced). In the damp grotto mosses and opposite-leaved golden-saxifrage thrive.

Belsay Hall and Castle are part of the English Heritage estate and close to Belsay village [NZ085784]. There is a bus service from Newcastle to Belsay.

Sandstones in the quarry garden

The Drake Stone

The Drake Stone is a very large and impressive block of sandstone resting on a craggy hillside. There are great views east and north over Harbottle and the Northumberland countryside.

The rock is Carboniferous and around 340 million years old. It is made of the same sandstone (the Fell Sandstone) that is the bedrock all around it and was probably detached from its original position by the action of ice. It isn't really a glacial erratic (a rock moved some distance from its bedrock by an ice sheet). As well as being huge the Stone shows the different layers and textures of the sandstone really well. It was once sand and pebbles carried down from the north by a very big river. The different layers and angles you can see in the sandstone show that this was a fast-flowing river with channels and underwater dunes.

There are many legends about the Drake Stone. People used to believe that if they touched it, it cured children of all illnesses and that if you spent the night beside it you would never leave.

Harbottle Moor, owned by Forestry England, is a nature reserve of Northumberland Wildlife Trust. On the way up to the Drake Stone you pass by drifts of bog myrtle. The shrub is aromatic, and was used to repel fleas and moths, gave a yellow dye, and flavoured beer. The Drake Stone is set in upland heather heath, with bilberry on the steepest slopes and peat bogs in the depressions.

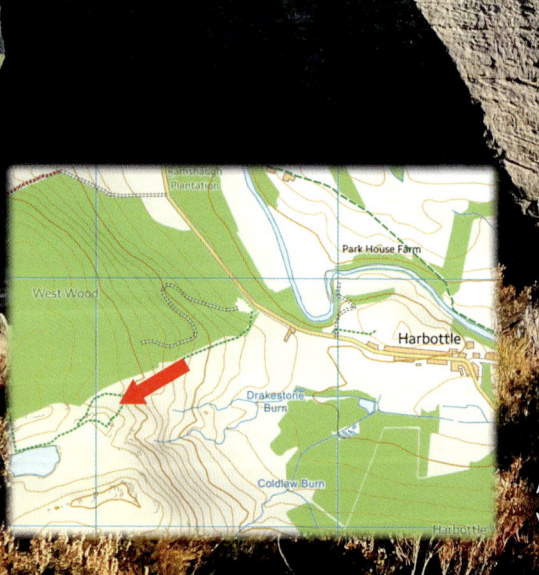

A 1 kilometre walk west of the village of Harbottle [NT920044].

Northern face of the Drake Stone

Green Carts Quarry

Green Carts is an old limestone quarry in a clearing in the middle of a sitka spruce forest. Lying on the floor of the quarry are thousands of fossils of large sea shells, a brachiopod called Gigantoproductus; they reach 15 cm across! Also there are fossils of corals and other sea creatures, all weathering out of limestone rock in the middle of Wark forest.

These rocks are about 335 million years old. The rock is called the Fourlaws Limestone and it's from the Carboniferous period. Brachiopods have been around more for than 550 million years and there are a few species still living today. Our fossil brachiopods lived attached to a rocky sea bed.

Limestones like this are made of calcium carbonate from the shells and skeletons of billions of sea creatures, including brachiopods, together with carbonate mud. The environment at the time was a warm shallow sub-tropical sea.

The floor of the quarry has a carpet of glaucous sedge and mouse ear hawkweed with fairy flax and bird's foot trefoil. There are lots of tiny spruce seedlings as well. In late summer see if you can find the spikes of autumn gentian in the southeast corner of the quarry. Roe deer are common and you will hear great tits and coal tits. If you are very, very lucky you may see a goshawk.

The quarry is in the southern part of Wark Forest [NY773715]. Finding it is a challenge! It is a long walk from the south – take the Pennine Way from just west of Housesteads. Or you could mountain bike in from the end of the tarmac road to Scotchcoultard.

Gigantoproductus fossils in limestone

Jesmond Dene

Jesmond Dene is a wooded valley which carries the Ouseburn. It is now a park within the city, but it hasn't always been like this. As well as corn and flint mills there used to be quarries in the thick sandstone that forms the sides of the Dene. Coal was mined here too and there were drifts and shafts in the valley side.

By 1862 much of the land in the Dene had been bought up by Lord Armstrong – he of Cragside and the armaments factories. After a lot of blasting of sandstone to create picturesque waterfalls and divert the Ouseburn, plus ornamental tree planting, he gifted it back to the people of the city in 1883.

The sandstone and coal are about 315 million years old from the Carboniferous period, but the valley was cut as the last ice sheet melted 15,000 years ago. Its steep sides and gorge-like shape are evidence that Jesmond Dene was eroded by a significantly more powerful flow of water than the stream that runs along it now. Probably the enormous quantities of water that drained from glacial lakes and melting ice sheets.

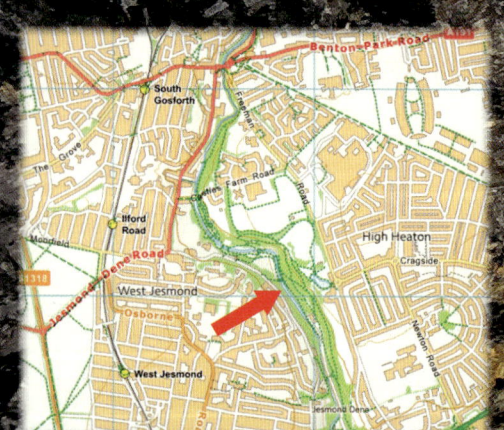

1 kilometre northeast of Newcastle city centre, easily reachable by foot, bus or metro [NZ262664].

Waterfall over sandstone blocks in Jesmond Dene

Kielder Stone

On the border between England and Scotland is a huge block of Fell Sandstone that sits in the midst of wild heather moors. In England it's the Kielder Stone, in Scotland it's the Stane. It's a long way from any road and a real challenge to reach.

The layers (beds) in the Stone vary in grainsize and show how sand was originally carried down by a large river whose flow rate was constantly changing. The Stone is a massive piece of sandstone that has been left exposed after millions of years of erosion, particularly by ice during the last glaciation. It is probably not an erratic (a rock transported by glacial ice) as it has not moved very far from its original bedrock position. The Fell Sandstone is Carboniferous and around 340 million years old. The ice sheets that eroded and may have detached the stone were active around 20,000 years ago.

As with all very large stones the Kielder Stone has many legends and stories associated with it. One, noted by Sir Walter Scott, says that if you go around it withershins (anti-clockwise) three times it will bring bad luck. It is said to be a place where people would meet and families would drop letters during the times of the Border Reivers.

Growing on the Kielder Stone are cowberry, bilberry and crowberry, and on the rock faces, because of the pure air and high humidity, shaggy lichen species. These are declining and rare elsewhere because of air pollution. You may see feral goats in the vicinity.

About a 5 kilometre hike from Deadwater, or a 10 kilometre hike north from Kielder Castle [NT636005], both over rough terrain.

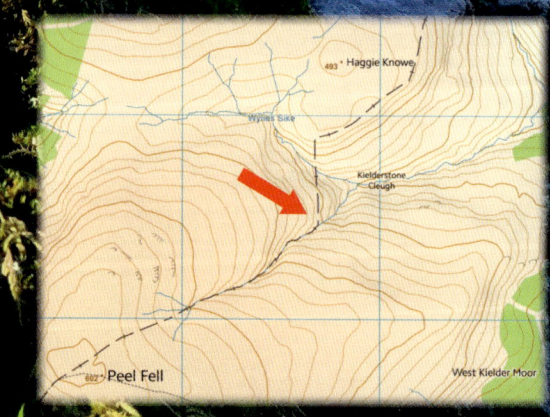

View northeast of the Kielder Stone

Queen's Crags

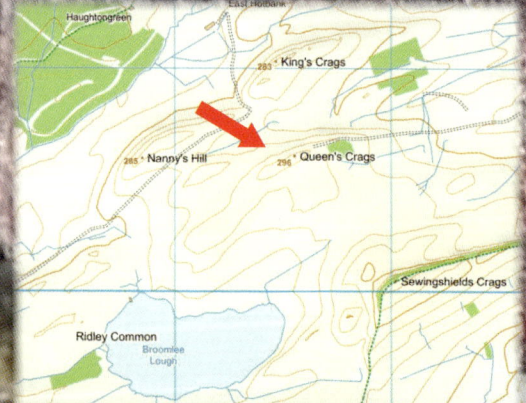

At the western end of the Crags is a very large angular sandstone boulder that has become detached from the rock face just to the south of it. The boulder looks like a giant tooth, or rabbit's ear and you can see it from miles away.

The sandstone is Carboniferous and about 320 million years old. The separation of the boulder from the crag probably happened as the last ice sheet melted away 15,000 years ago.

Just a few metres east of the boulder, along the crag and under an overhang, there is a Roman inscription. Two Centurions and an Optio (an "under-centurion") carved their names in the rock, possibly while sheltering from the weather.

Sandstones produce acidic soils. Bracken is growing on the top of the crags, and in front is a mixture of heather and purple moor-grass. The grass is our only deciduous one, losing leaves and blowing around in winter.

North of Hadrian's Wall. You can reach it by hiking from Housesteads Visitor Centre. [NY794705]. The Hadrian's Wall bus service could be used to reach Housesteads.

The sandstone boulder at Queen's Crags

Simonside

This bold sandstone escarpment is one of the iconic gems of the landscape of Northumberland. Its crags and towers gaze north over Coquetdale and towards the Cheviots.

335 million years ago these sandstones were sand and pebbles in an enormous river meandering across a broad flood plain from uplands in the distant north. If you can conjure up an image of the braided channels of the modern River Brahmaputra in Bangladesh you won't be far off. Erosion over millions of years, and especially during the last glaciation, has left this hard Fell Sandstone ridge standing loftily above the surrounding countryside.

Some believe Simonside to have been a sacred mountain to our ancestors, who built the cairns and tumuli and carved the rock art at nearby Lordenshaw. But regardless of whether Simonside was sacred, it's safe to say those early people understood the power and influence of these rocks on our material and spiritual lives more than most of us do today.

This mosaic of wet and dry moorland is now mainly covered in heather but there's bilberry and cross leaved heath too and in the bogs are hare's-tail cotton grass and the insect eating plant round-leaved sundew. Waders such as curlews and golden plover breed here, as do kestrels and ravens and the ubiquitous red grouse – this is a landscape managed for the purpose of grouse shooting.

South of Rothbury. You can park at Lordenshaw [NZ052988] and take a look at the rock art there too.

Looking north across the Simonside sandstone escarpment

Sugar Sands Bay

This bay is a fantastic place to see the different layers of rock that were once sand, peat, mud and coral reefs and the fossils of the animals and plants that lived here 325 million years ago in the Carboniferous period.

All these different layers of sandstone, shale, limestone and coal tell us that the environment repeatedly changed from fast flowing rivers, to swampy forests, coastal lagoons, to coral seas. They also show that sea level kept rising and falling many times; probably because the Earth's polar ice caps were melting and then expanding. So we know that climate change has happened before – many times.

As well as fossils of sea shells (brachiopods) you can find fossil corals, sea lilies (crinoids) and the burrows of ancient animals.

The Bay is well-known botanically as a site for seaside centaury, and several sedges including distant sedge, which is not common in Northumberland. In the autumn Agrimony can be found and several salt loving plants such as orache.

Near Low Stead Farm [NU258163], 2 kilometres east-northeast of Longhoughton. Walk north along the foreshore.

Alternating sedimentary rock layers in Sugar Sands Bay cliffs

Tynemouth Castle and Priory

The youngest bits of sedimentary bedrock in the whole of old Northumberland are here, in the cliffs under the castle and priory. Sitting on top of Carboniferous sandstones, is a yellow sandstone and on top of that a pale rock called a dolostone (like a limestone but with magnesium as well as calcium).

The yellow sandstone and the dolostone are 260 million years old, from the Permian period of Earth's history. They tell a tale of a stony desert that was covered by wind-blown sand dunes (the yellow sandstone). Then the sea invaded the land and deposited silt and mud (which has fish fossils, but you can't see the shale very well here) and then the magnesium- and calcium-rich dolostone. The sea, called the Zechstein Sea, was very salty, a bit like the Dead Sea today.

Beside the steps of the north pier, where it meets the cliff, look down and east; a 3 metre band of darker grey rock heads out to sea, parallel to the pier. This is the Tynemouth Dyke – an igneous rock called dolerite. It is only 60 million years old and one of the youngest rocks in Northumberland; a red-hot visitor from volcanos near the Isle of Mull.

The south-facing cliffs and slopes above the Haven go yellow in spring from wild cabbages and alexanders. Both are ancient introductions and probably date back to when the monks were in residence. Wild forms of wallflower are also conspicuous. These slopes can be a good place to find migrant birds, particularly in autumn. The seaward facing cliffs have nesting fulmars, which are usually in residence from January onwards.

Easily reachable by public transport. Walk to the north pier from Tynemouth Castle, or you can view the cliffs from King Edward's Bay [NZ374693].

TYNEMOUTH

View south of Tynemouth Castle and Priory

Coquet Valley

If you enjoy treasure-hunting then you might want to explore the shingle banks of the River Coquet between Alwinton and Blindburn and search for agates.

These beautiful small stones, which geologists call amygdales, contain concentric bands of quartz of different colours. Because of their attractiveness they are often used in jewelry.

The agates originated in cavities in the Cheviot volcanic rocks. These cavities were once gas bubbles, vesicles, that were filled with hot mineral-rich fluids which then slowly cooled and crystallised. The agates then weathered out of the rock and were carried downstream by the river. The volcanic lavas that the agates were once part of are about 400 million years old and from the Devonian period of Earth's history.

While the Coquet valley is a good place to look for agates, you could search almost any valley or stream that drains the Cheviot volcanic rocks.

Dippers, common sandpipers and grey herons are common, and oystercatchers breed on the shingle. The volcanic lavas are locally quite rich in calcium, so that lime-loving plants like hairy rock-cress and quaking grass occur on rocky outcrops by the river. You may see feral goats on the nearby hills.

Anywhere upstream of Alwinton [NT913062] where there is public access to the shingle banks beside the river.

10 mm

Agates from the shingle banks of the River Coquet

Cuddy's Crags

This is one of the best views of the Roman Wall marching eastwards. It is a fantastic and extensive outlook on the Whin Sill, the hard igneous rock, called dolerite, which forms steep cliffs facing north.

The Whin Sill was inserted as a sheet into local Carboniferous rocks 295 million years ago. But there have been millions of years of erosion since which has left the hard dolerite standing proud. The rock was once billions of tons of molten magma injected from deep in the Earth in between the layers of Carboniferous limestones, sandstones and shales. Then it cooled and solidified and contracted forming prominent vertical cracks and fissures.

All the ups and downs of Hadrian's Wall and the National Trail along it, are a result of ice and meltwater during the last Ice Age eroding through the Whin Sill ridge. So you can blame geology for those tired legs after you have finished your day's walking!

The Wall is home to the common lizard, stoat, nesting wheatear, maidenhair spleenwort and wall rue. Thyme and lady's bedstraw grow both on the Wall and alongside it. Skylark, meadow pipit and raven soar. Further west rock-rose delights. golden-scaled male-fern clothes the Crag's northern face.

Cuddy's Crags is 750 metres west of Housesteads Fort [NY783686]. The Hadrian's Wall bus service stops here and there is a large car park at the Housesteads Visitor Centre.

The Roman Wall looking east over Cuddy's Crags

Cullernose Point

Just over 1 kilometre walk south of Craster [NU260187].

Cullernose Point is a very prominent, rugged cliff tapering out to sea. It is composed of dolerite, the extremely hard rock that forms the Whin Sill. In the bay to the south are gently folded limestones and sandstones that have been cut by a separate vertical blade of Whin.

The Point is the best place on the Northumberland coast to see the Whin Sill's very distinctive vertical fractures, which are characteristic of many igneous rocks. It's called columnar jointing and the classic examples in the British Isles are on the Island of Staffa and the Giants Causeway, in Northern Ireland, where the igneous rocks are basalt lava flows.

295 million years ago when this part of the Earth's crust was being stretched apart, billions of tons of molten magma was injected as a sheet from deep in the Earth into other rocks across many miles of northern England.

The salt-tolerant fern sea spleenwort grows on the cliffs, and fulmars nest on ledges. The Point is a good place to watch birds out at sea – gannets are frequently seen.

The Whin Sill at Cullernose Point with folded limestones in foreground

Great Standrop

Great Standrop is a tor: a castle of granite rock standing proud after the rock around it has been weathered and eroded. The views from the top of the tor are panoramic.

The granite is 400 million years old and was formed in the Devonian Period. But the weathering of the rock and its formation as a tor has taken millions of years since, and particularly during the Ice Age – the last 2.6 million years. Tors form because their rock is more resistant to erosion than the immediately surrounding ones. Usually this is because the joints (cracks) which let in the soil water that rots the rock, or breaks it up by frost action, are further apart. Our tors, unlike those on Dartmoor, have been battered by glaciers, so are not quite as prominent.

The moorlands are a mosaic of heather, acidic grassland, bracken and – on higher ground - blanket bog. They are managed for grouse and sheep, and the patchwork burning of the heather is to regenerate young, more nutritious shoots. The blanket bog also has heather, mixed with hair's-tail cotton grass and there are also the bramble-like leaves of cloudberry – a higher-altitude species with orange berries later in the summer. Meadow pipits are the most common birds, and you may see common lizards and adders.

At the head of the Breamish Valley. Follow the road as far as Hartside Farm. The walk to Great Standrop is a 9 km round trip [NT943180].

Great Standrop granite tor from the southeast

Holy Island

On this small island, also known as Lindisfarne, there are lots of special rocky places to visit. But, because it is fundamental to the existence and heritage of the island, the focus here will be the Whin Dyke.

Like the Whin Sill suite of rocks, of which it is part, the Whin Dyke is a hard igneous rock called dolerite, injected vertically into existing Carboniferous sedimentary rocks. Where the Dyke cuts like a black blade across the foreshore, defining the south side of the Heugh, look out for fossils of crinoids and brachiopods in the adjacent limestone.

Near the castle you can also visit some of the largest limekilns in Northumberland and then walk north to see the limestone which they quarried near Nessend and burnt in the kilns. At Snipe Point there are stunning rock troughs and domes where the rocks have been folded.

While the igneous and sedimentary rocks are the bedrock of Lindisfarne, the overall shape of the island today is very much influenced by much more recent wind-blown sand dunes which are less than 6000 years old.

The Whin Dyke is in many ways the physical foundation of Lindisfarne's heritage. Without it there would have been no island for St Cuthbert's Hermitage, no settlement of the Heugh, and no site for the castle…… and a lot less visitors today.

The dolerite is hard, so weathers and releases nutrients slowly, and on its shallow soils in dry seasons the vegetation can be physically-stressed. On the Heugh is a distinctive set of species, with common rock-rose and several uncommon clovers including slender trefoil and rough clover.

Just off the coast, south of Berwick upon Tweed and north of Alnwick [NU136418]. You can only reach it when the tides are right.

Holy Island Castle built on the Whin Dyke

Housey Crags

Housey Crags is another spectacular Cheviot tor but this natural rock castle is not made of granite, it was once lava from a volcano. It sticks out on a plateau on the hillside and provides great views of The Cheviot, Northumberland's highest mountain.

The volcanos in this area erupted about 400 million years ago and their lava covered much of northern Northumberland. When they were erupting Britain was 30 degrees south of the Equator, not 55 degrees north like now. These crags are made of rocks that exploded violently as molten magma from volcanic vents. The rocks are called andesites, but shortly after they erupted they were heated up and hardened by the molten Cheviot granite; they are now a metamorphic rock called a hornfels.

Millions of years of erosion, especially during recent Ice Ages, has left them poking out of the ground.

Surrounding the Crags the grass and heather moorland is managed for red grouse. Skylark, meadow pipit and curlew are common in the spring and summer and you might see a common lizard. The Crags have good lichen cover and a range of mosses.

In the Harthope Valley, southwest of Wooler [NT957217]. It's a 2 kilometre walk there and back from the road.

Housey Crags from the southwest

Limestone Corner

Limestone Corner is not limestone, it's Whin Sill dolerite. This is part of the ditch that the Romans dug immediately north of Hadrian's Wall. Here they did not succeed, the hard igneous rock defeated them.

You can see their attempt to create the ditch; broken boulders are scattered around and in one large block you can see the chisel marks the Romans made to try and split it. Curiously their colleagues digging the Vallum (the behind-the-Wall ditch), in the same rock about 100 metres further south, managed to finish their job. The Wall and the ditches were constructed between 122 and 128 AD.

The Whin Sill is an igneous rock. 295 million years ago it was molten and injected into the surrounding Carboniferous sandstones, limestones and shales.

As you walk along the base of the quarried ditch look out for wild thyme and wood sage growing in cracks and on the shelves amongst the rocks. The dandelion-like flowers of mouse-ear hawkweed are also common here. At the eastern end of the channel see if you can find the tufts of parsley fern, uncommon in Northumberland, growing in shady crevices.

Immediately north of the B3618 Military Road [NY874716]. 100 metres from a small parking pull-in.

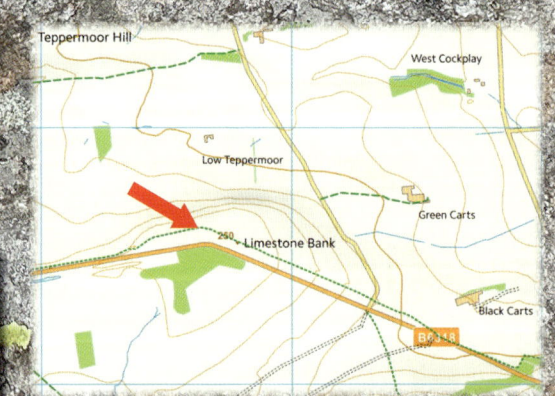

The partially complete Roman ditch through the Whin Sill at Limestone Corner

Linhope Spout

Linhope Spout is a spectacular 18 metre waterfall cascading over Cheviot granite rocks. The granite is 400 million years old and was formed in the Devonian Period.

The Cheviot Granite was once molten rock, deep in the Earth, with a temperature of more than 1200 degrees. It was so hot that it baked the surrounding volcanic rocks. Such a large volume of molten magma and at such great depths cools and solidifies slowly. This allows the minerals within the magma (quartz, feldspar and biotite) to form large interlocking crystals that can be seen by the naked eye. The granite is over 60 square kilometres at the ground surface but at a depth of 4 kilometres it is over 300 square kilometres in area. It was once investigated as a possible deep source of geothermal energy.

Ash, elder and birch grow by the waterfall, with willow warblers. The Spout has abundant mosses, great wood rush and opposite-leaved golden saxifrage; drier ground has common dog violet.

At the head of the Breamish Valley. Follow the road as far as Hartside Farm. The walk to the Spout is a 5 km round trip [NT958171].

Linhope Spout waterfall cascading over Cheviot granite

Lumsden Law

When you drive up to Scotland on the A68, just after Catcleugh reservoir and before Carter Bar, there is a big flat-topped hill to the east. It's called Lumsden Law. Geologists think it could be a volcanic 'plug'; that is lava that solidified in the neck of a volcano.

This volcano erupted about 345 million years ago, so it's Carboniferous and younger and different in composition to the Devonian lavas of the Cheviots. While there are lots of volcanic rocks of this age in Scotland there are very few like them south of the border. Lumsden Law was formed in a similar way and around the same time as the rock that Edinburgh Castle stands on.

During this period of the Carboniferous in southern Scotland and here at Lumsden Law, there was a lot of volcanic activity – the molten magma cooled to produce a rock called a basalt. You can see the rock close-up in an old quarry on the Law's western flank. A broken face will show some quite large crystals in a darker fine-grained matrix, a texture called porphyritic.

The basalt in the quarry has a rich covering of several different lichens. From the Law there is a fine view westwards across the A68 into Northumberland Wildlife Trust's largest nature reserve, Whitelee Moor. The higher ground of Whitelee has some of the best blanket bog in England and below is upland heather heath.

800 metres, northeast of Ramshope Lodge on the A68 [NT722052]. It is a rough hike to the top.

Lumsden Law basalt rock with lichens

Walltown

Walltown is a huge, restored old quarry in the Whin Sill. The quarry opened in 1876 and closed in 1976. The Whin Sill (whinstone to quarrymen, dolerite to geologists) is so hard that they had to use explosives to blast it from the rock face.

Its hardness meant it was perfect for roadstone chippings used in tarmac and a century ago, for setts for cobbling streets. The Whin Sill was once molten magma; it was injected from deep in the Earth in between the layers of Carboniferous limestones, sandstones and shales and then it cooled, solidified and contracted forming its prominent vertical cracks and fissures. Millions of years of erosion since left the hard dolerite rock ridge standing proud...... until it, and the section of Hadrian's Wall that ran along it, was quarried away.

The enormous hole that was quarried out was infilled with compacted stony glacial clay as part of its restoration. Slowly nature is returning and there is now a wide assortment of plants on the varied soils and rock faces of the quarry. Look out for common rock-rose on thin whin soils. On the quarry floor are common spotted-orchid and northern marsh-orchid (and their vigorous hybrid), with numerous sedge species.

North of the B6318 east of Greenhead [NY669660]. The Hadrian's Wall bus stops at Walltown.

View east over Walltown quarry

Berwick-upon-Tweed

On the foreshore just east and north of Berwick-upon-Tweed is Ladies Skerrs. When the tide falls it reveals a dramatic series of rocky whorls and swirls; concentric rings and curves. They look like a bomb crater, or as if someone has sliced through pudding bowls stacked upside-down. At Bucket Rocks, beside it, the bowls are the right way up.

The rocks are limestones, sandstones, shales and thin coals deposited in Carboniferous times 330 million years ago. While it doesn't seem possible, even brittle substances like rocks can be bent and folded if the process is slow and steady and done very, very gradually.

The rocks were bent and folded when Britain and northern Europe were part of the building of an enormous mountain chain about 300 million years ago. The sea has then eroded away the top parts of the rock "bowls".

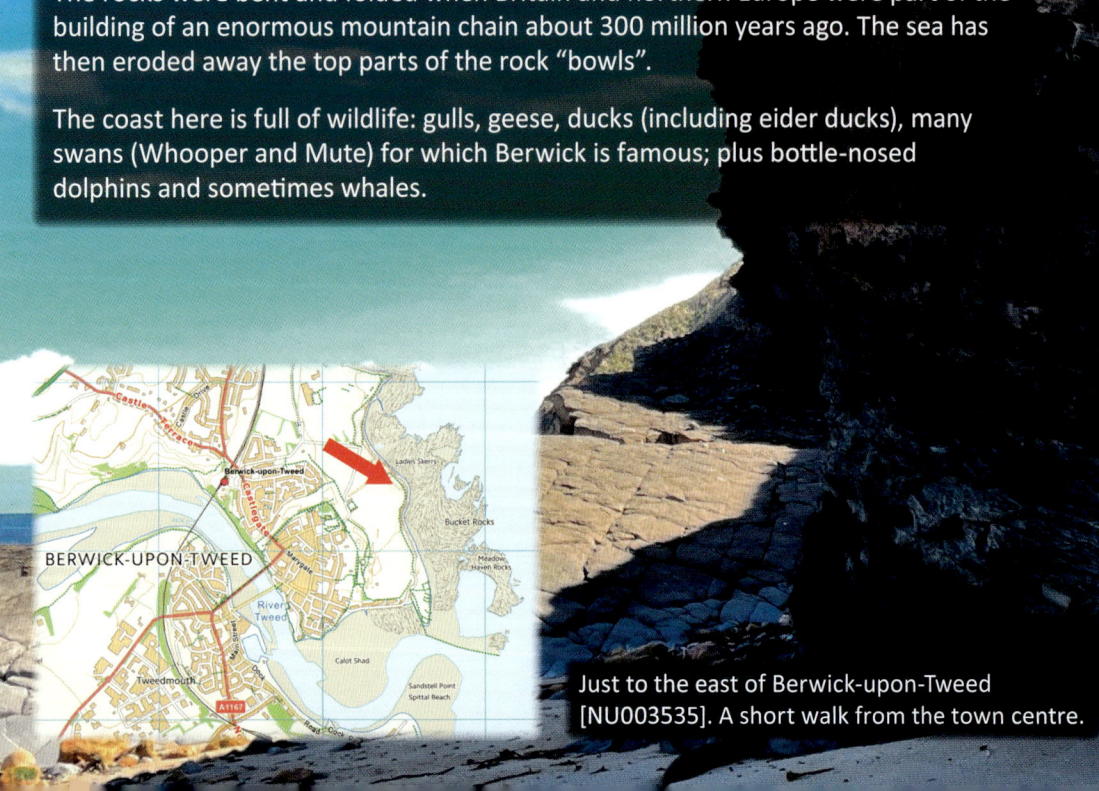

The coast here is full of wildlife: gulls, geese, ducks (including eider ducks), many swans (Whooper and Mute) for which Berwick is famous; plus bottle-nosed dolphins and sometimes whales.

Just to the east of Berwick-upon-Tweed [NU003535]. A short walk from the town centre.

Sandstones, limestones and shales folded into a dome and eroded by the sea at Ladies Skerrs

Coquet Head

Right next to our border with Scotland are the oldest rocks in Northumberland. Perhaps the best place to see them is at the head of the river Coquet, near a farm called Makendon. In a little river cliff, beside a bridge, are shattered and broken grey-green rocks at a steep angle.

These shales were once mud on a deep ocean floor. With them are rocks called greywacke or turbidites. They were mud and sand that slid down the continental slope so quickly they didn't have time to settle out and are all mixed up. They are very old and so have been affected by many pressures within the Earth; that's why they are at such strange angles and are so broken and fractured.

The rocks are Silurian in age, that's 425 million years ago. Northumberland doesn't have any rocks older than this and they only occur around here in the extreme northwest of the county, stretching across the A68 to Northumberland Wildlife Trust's biggest reserve at Whitelee Moor. In places the rocks contain very primitive marine fossils called graptolites and acritarchs.

Just a short distance up the road there is some great archaeology - the Roman camps and fort at Chew Green, Dere Street - the Roman Road, and a medieval village called Kemylpethe, that was probably used as a stop-over by Scottish cattle drovers.

You may see the herd of wild goats (escapees from domestication, perhaps centuries or millennia ago), and in spring and early summer hear the sound of curlews and skylarks.

A long drive up a single track road to the head of the River Coquet, past Blindburn [NT806096] and beside a bridge over the stream.

Broken ancient rocks near Makendon, once deep ocean sediments

Crindledykes Quarry

Crindledykes is an old quarry which used to work a thick layer of limestone called the Great Limestone. The rocks in it have been bent and buckled into 'folds' by earthquakes.

The limestone dates from Carboniferous times, 330 million years ago. The limestone is here because at that time all of northern England was just 5 degrees north of the Equator and was covered by shallow sub-tropical seas. It is the billions of shells, corals and small animals and plants that lived in those seas that make all limestones. These rocks were folded when Britain and Europe were part of the building of an enormous mountain chain 300 million years ago.

The quarry is here because lime is valuable for making acid soils more fertile and it is an essential ingredient of cement. Just west of the quarry is a restored limekiln where they used to burn the limestone to make lime.

Limestone bedrock produces a very particular set of plants. Here they include: fairy flax, small scabious, hairy rock-cress, salad burnet, cowslip, hoary plantain and quaking grass.

The quarry is now a Northumberland Wildlife Trust nature reserve and is half a kilometre north of the road known as the Stanegate, north of Bardon Mill [NY784672].

Buckled limestones in an old quarry at Crindledykes

Cullercoats Bay

On the foreshore at Cullercoats Bay is a big geological fault; a place where rocks have been broken and dislocated by enormous forces in the Earth.

Two sections of rock that are millions of years different in age have been left lying next to each other. To the south are dark grey shales of the Carboniferous; to the north is the younger, Permian, yellow sandstone. But this break in the Earth's crust has a longer history and has probably also been a weak point for a long time.

Geologists have traced this fault and a broad line of other changes in the rocks deep in the crust using geophysics. They concluded that they are part of the "Iapetus Suture Zone": a zone of weakness caused because two ancient continents collided hereabouts when the ancient Iapetus Ocean closed about 420 million years ago.

Coal miners regularly found the fault as they extracted over 20 coal seams below southeast Northumberland. They came to recognize that it would dislocate the coal seams by around 150 vertical metres, so they named it the 90 Fathom Dyke.

You might see purple sandpipers feeding near the water's edge, and it's a good place for watching birds out to sea.

The fault is to the south of Cullercoats Bay and its South Pier [NZ365710]. It's easy to reach by bus or car.

Relative direction of rock movement

The 90 Fathom "Dyke" (fault) at Cullercoats Bay

Howick Bay

In the cliff backing the Bay is a very large fracture in the rocks. It is the Howick Fault. It can be clearly seen in the cliff and can be traced out east across the foreshore. The rocks to the south have 'dropped' more than 200 metres in comparison to the rocks to the north of the fault.

The rocks either side of the fault – limestones, shales and sandstones - are around 330 million years old. The rocks to the south of the fault are marginally younger than those to the north. The fault also had a thin blade of Whin Sill dolerite intruded into it 295 million years ago. The stresses that caused the fault, (just like earthquakes) and allowed the intrusion of the Whin dyke, were part of a long period of mountain building in Britain that started 300 million years ago.

There are many fossils of shells and crinoids in the limestones both north and south of the fault. The sandstones have fossils too; of bits of Carboniferous trees. About 200 metres south of the fault, on the sandstone foreshore where the cliff turns eastward, geologists found some of the oldest footprints in Britain – a four-legged amphibian. The sea has all but eroded them away now.

You might see eider ducks, as well as purple sandpipers, feeding near the water's edge, and it's a good place for watching birds out to sea.

400 metres northeast of Howick [NU259180]. There is a small roadside parking area just above the Bay.

Relative direction of rock movement

The Howick Fault in the cliff and running across the foreshore

Saltpan Rocks

On the foreshore, near Scremerston, limestone rock has been bent into one incredible tight fold and several arches (anticlines).

In the limestone, which has been polished by the sea, there are many thousands of fossils of ancient sea shells, called brachiopods, and corals.

The limestones, which alternate with sandstones, shales and thin coals, were deposited in a sub-tropical clear, coral sea in Carboniferous times 330 million years ago. The rocks were bent and folded when Britain and northern Europe were part of the building of an enormous mountain chain which began about 300 million years ago.

The changing rock sequence – limestones, followed by shales (mudstones), then sandstones, seatearths (fossil soils), and thin coals – is repeated many times. Geologists have concluded that this tells a story of rising and falling sea level as the Earth's ice caps grew and then melted millions of years ago. In many ways that's like the period of Earth's history we live in today.

The sand dunes nearby are calcareous because of the many shell fragments. Plants include purple milk-vetch, burnet rose, viper's bugloss and bloody crane's-bill (including a pale pink variety). On the backshore is Scots lovage, which reaches its southern-most east coast limit in north Northumberland. Shorebirds include purple sandpiper, turnstone and oystercatcher.

Just to the east of Scremerston you can park beside the coast road at Cocklawburn Beach and walk north. [NU024493].

Folded limestone on the foreshore at Saltpan Rocks

Beldon Cleugh

Beldon Burn flows east down from the Pennine hills. About 5 kilometres west of Blanchland it is joined from the north by a large, steep sided, valley which has no stream running in it. During the last Ice Age this channel was cut and used by torrents of glacial meltwater flowing under an ice sheet.

Geologists know it was cut by a river under pressure under the ice because it has a "humped profile" – its course goes up and over a col. Beldon Cleugh is even more remarkable than it looks because there are at least 9 metres of peat infilling its floor.

During the last glaciation, only 20,000 years ago, all of northern Britain was covered by a sheet of ice, up to a kilometre thick in places. Ice that thick completely changed the landscape and blocked normal valleys and drainage routes. Water beneath the ice cut new channels, often completely ignoring the pre-existing topography.

Bog vegetation on the peat infill consists mainly of heather and hare's-tail cottongrass. Bracken grows on the steep sides. The surrounding heather moorland is managed for red grouse.

Beldon Cleugh is a 5 kilometre hike west from Blanchland [NY917505].

A very accurate model of the ground surface showing the Beldon Cleugh channel.

LiDAR Composite Digital Terrain Model at 1m spatial resolution produced by the Environment Agency. Public sector information licensed under the Open Government Licence v3.0

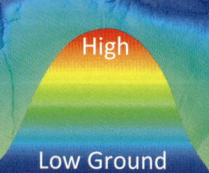

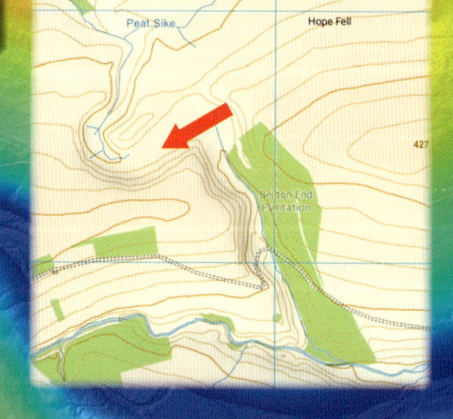

View south of the glacial meltwater channel at Beldon Cleugh

Bradford Kame

Bradford Kame (or Kaimes) is a striking, sinuous, steep-sided ridge, called an esker, winding north-south across the landscape.

It's about 15 metres high and more than 3 kilometres long and is made mostly of sand and gravel. It's a very important UK scientific site – a Site of Special Scientific Interest - and is one of very few geological features in Britain explicitly named on an Ordnance Survey map.

The ridge was formed around 15,000-20,000 years ago during the last glaciation of Britain. In the last Ice Age all of Scotland and much of northern England, was covered by an ice sheet that was in places almost 1,000 metres thick. When the climate warmed and the ice sheet began melting it left a covering of clay and stones over most of the land. But in some places, where rivers carried meltwater and debris from under, on top of and within the glacier, it left sand and gravel. Sometimes this was deposited in tunnels under the ice (eskers) and sometimes as deltas and fans in lakes next to the ice.

Our planet is currently in a warm period – an interglacial – and were it not for human impact on our climate, in a few tens of thousands of years the glaciers would have certainly returned. We may have deferred the next ice age but there is a price to pay - very different and more immediate climate extremes.

The ridge is difficult to farm so has been left largely to nature, with a tangle of bramble, gorse and bracken, useful for birds such as linnet and song thrush to feed and nest in. But you may also see buzzards, sedge warblers, skylarks, willow warblers, whitethroats and spotted flycatchers.

Bradford Kame is between Bamburgh and Lucker [NU162320].

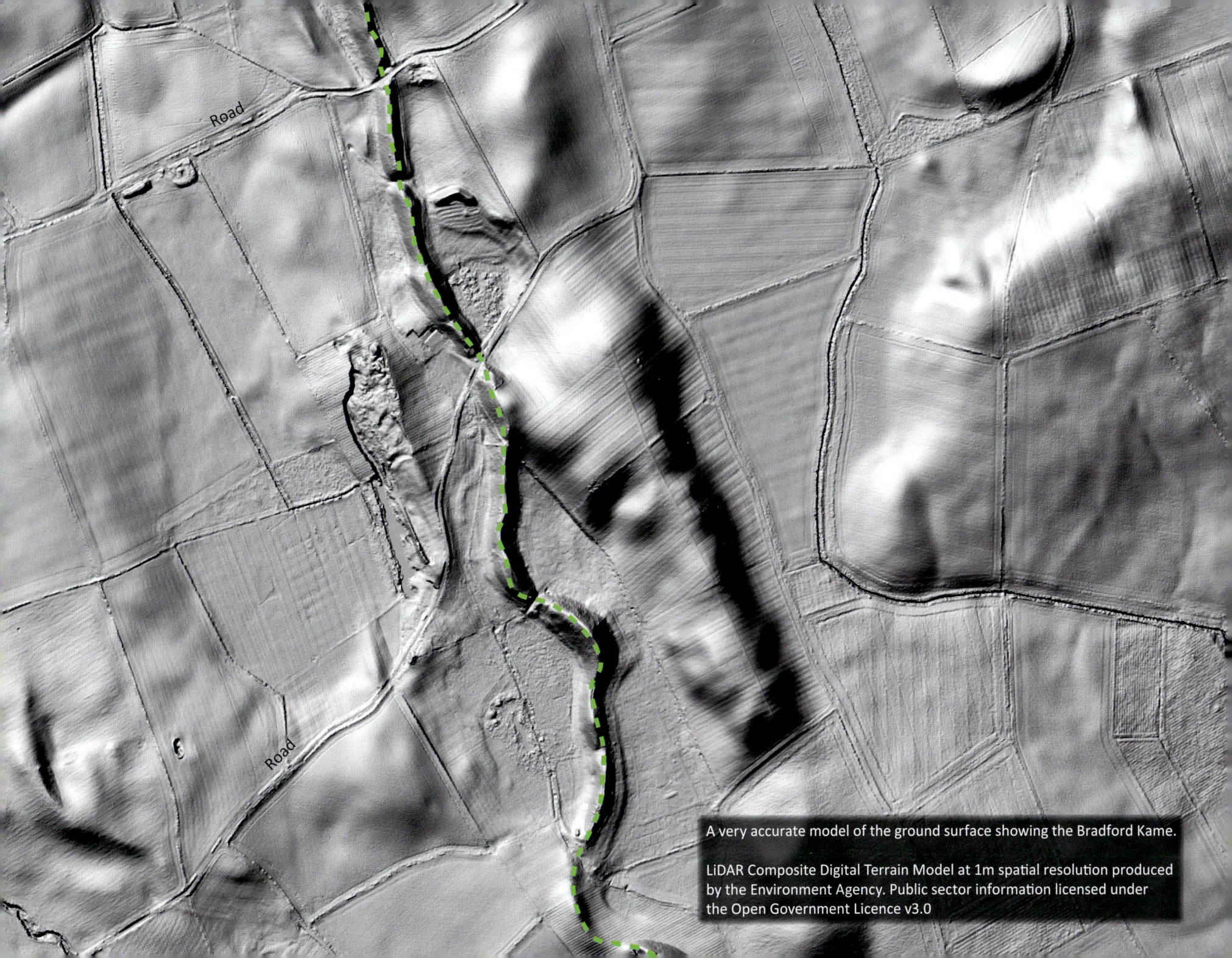

A very accurate model of the ground surface showing the Bradford Kame.

LiDAR Composite Digital Terrain Model at 1m spatial resolution produced by the Environment Agency. Public sector information licensed under the Open Government Licence v3.0

Broomlee Lough

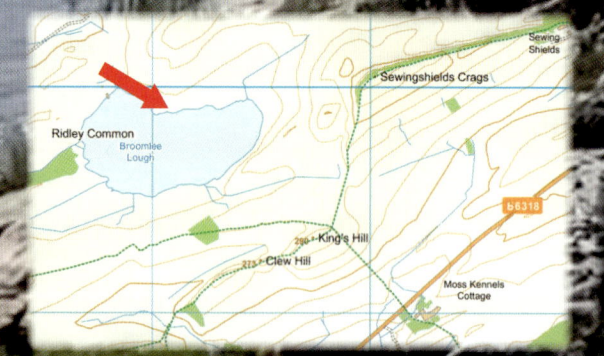

The pavement is on the northern shore of Broomlee Lough, which is north of Hadrian's Wall. You can reach it by walking from Housesteads visitor centre. [NY790698]. The Hadrian's Wall bus service stops at Housesteads.

On the northern shore of Broomlee Lough is a limestone pavement: a natural platform of bare limestone rock cut by fissures, like a very rough set of paving slabs.

The limestone is Carboniferous and called the Oxford Limestone. It is over 320 million years old but the "pavement" feature was mostly created after the last ice sheet left 15,000 years ago. This is a very rare feature for Northumberland. There are many limestone pavements in the Yorkshire Dales and in south Cumbria but not in our county.

It happens when an ice sheet scrapes off all the soil and loose rock and leaves the limestone exposed. That leaves the rock open to the elements. Rain absorbs carbon dioxide from the atmosphere and the weak carbonic acid eats into the joints of the limestone, which is mostly made up of calcium carbonate – creating this natural pavement. The blocks are called clints and the fissures are called grikes.

This fragment of rare limestone pavement supports lime-loving plants such as small scabious, salad burnet, common rock-rose, fairy flax and mouse-ear hawkweed.

View west of Broomlee Lough from near Sewingshields Crags

Down Hill

The B6318, Military Road, is usually a very straight east-west road, which for much of its course follows the line of Hadrian's Wall and its defences.

At Down Hill, however, the road does a chicane around a small hill, as does the Roman ditch, the Vallum, which diverts round it on the south side.

This is a very unusual small hill because it is a giant erratic, a glacial raft of limestone rock, a megablock, that has been plucked from its original position by an ice sheet and carried at least 2 kilometres to rest where it is now.

The limestone (called the Great Limestone) is Carboniferous and is over 320 million years old but as a slab of rock carried by an ice sheet it is possibly only around 20,000 years old and certainly less than 2.4 million. The ice sheet that carried this rock was moving quite quickly from Scotland and Cumbria across Northumberland and as it did so it not only carried lots of rock and debris, within it and frozen to its base, but sometimes literally plucked large pieces of bedrock and took them downstream.

This is possibly one of the larger glacial rafts/erratics in England and because it is so large it puzzled geologists for a long time. It is big (approximately 600x300x15 metres, that's over 7 million tonnes) and has been quarried for limestone, including by the Romans. Usually erratic boulders are much smaller – a lot less than a cubic metre.

Growing on the old quarry spoil banks are lime-loving plants like salad burnet, mouse-ear hawkweed and quaking grass.

About 2 kilometres east of the Port Gate (Errington Arms) roundabout on the B6318. Walk back east along Hadrian's Wall Trail [NZ006684].

A very accurate model of the ground surface around the Down Hill giant erratic. The erratic is shaded green.

LiDAR Composite Digital Surface Model at 1m spatial resolution produced by the Environment Agency. Public sector information licensed under the Open Government Licence v3.0

The Down Hill giant erratic from the west

Druridge Bay

Between Hauxley and Cresswell is a beautiful 9 kilometre sandy beach and bay that's longer than any other on the Northumberland coast. It's backed by a set of sand dunes and at times when the sea scours the shoreline, it reveals peat and the stumps of ancient trees.

The beach is constantly moving and changing. Every day, with each tide, it has a different shape, profile and composition. So this sandy bay is the youngest "rock" in this book. The oldest sand dunes are less than 6,000 years old. The peat and the trees are around 7,000 years old and sometimes, if the sea erodes through the peat you will get a glimpse of glacial clay and stones underneath that is perhaps 15,000 - 20,000 years old.

The ancient trees – sub-fossils – were once part of an extensive forest and land area that went all the way across the North Sea to the Netherlands. You could have walked across from the continent – and ancient humans did!

The dunes are covered in marram grass but you will also find white clover, birds-foot trefoil and bloody cranes-bill. The Bay has several nature reserves near it run by Northumberland Wildlife Trust.

North of Cresswell and south of Hauxley [NZ273984]. There is parking on the road west of the bay.

Ancient tree stumps revealed by the tide at Druridge Bay

Echo Crags

On a remote hillside tucked away in the northwest of the county is a dramatic, weathered and worn sandstone escarpment with expansive views over north Northumberland's hills and valleys.

These rocks are part of the Fell Sandstone. They form crags and escarpments in a big arc around the Cheviot hills and across Northumberland. 335 million years ago, during early Carboniferous times, they were sand and pebbles in a meandering and braided river system flowing across a broad plain. The way the rocks look now, their shape and texture as pock-marked crags, is mostly the result of weathering during and after the last glaciation of northern Britain. As well as the natural rocks outcrops, with their characteristic holes and hollows, there is a small quarry where you can see sedimentary structures that show evidence of the fluvial origin of the sandstone.

The rock outcrops are notable for their rich upland lichen assemblage, including several species that are common in the Scottish highlands but very rarely found south of the border. The most important of these are Alectoria sarmentosa, also known from the Kielder Stone, and Platismatia norvegica at its only confirmed site in England. The Alectoria is one of the species that could be a relict of native pine forest, of which a fragment in neighbouring North Tynedale is a possible survival. The crags are set among bilberry and heather.

Echo Crags is northwest of Byrness on the A68 road to Scotland. You can park at Byrness village [NT742044]. It is a rough 3 kilometre hike from there.

Echo Crags looking north

Fallowfield

Just south of the Military Road is a flat outcrop of Carboniferous sandstone with little or no soil or grass covering it. It looks as if it has been scraped clean by a massive road-planing machine. This, and the scratches in it, make it a very special bit of that sandstone that has helped geologists understand what our landscape has experienced.

The flat surface of the sandstone and the scratches are evidence that the area was eroded by a thick fast moving ice sheet only 20,000 years ago. The scratches that glaciers and ice sheets make are called glacial striae. They show the direction of movement - generally west-north-west to east-south-east. Some of the scratches here appear to accentuate sedimentary features in the sandstone.

This ice sheet was moving from Scotland and Cumbria across Northumberland and as it did so it scraped, bulldozed and carried billions of tons of rock, clay and sand with it. Stones and pebbles at the base of the moving ice literally "sand-papered" the bedrock beneath and gouged grooves into it.

This is pasture land with only a thin soil but look for pineapple weed growing in the joints of the bedrock; it smells like pineapple.

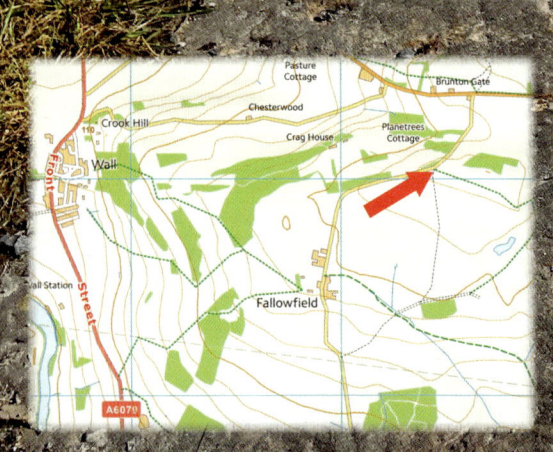

On a footpath just off a minor road south of the Military Road, the B6318 north of Hexham and east of Wall [NY935691].

Sandstone bedrock eroded and scratched by an ice sheet

Falstone Moss

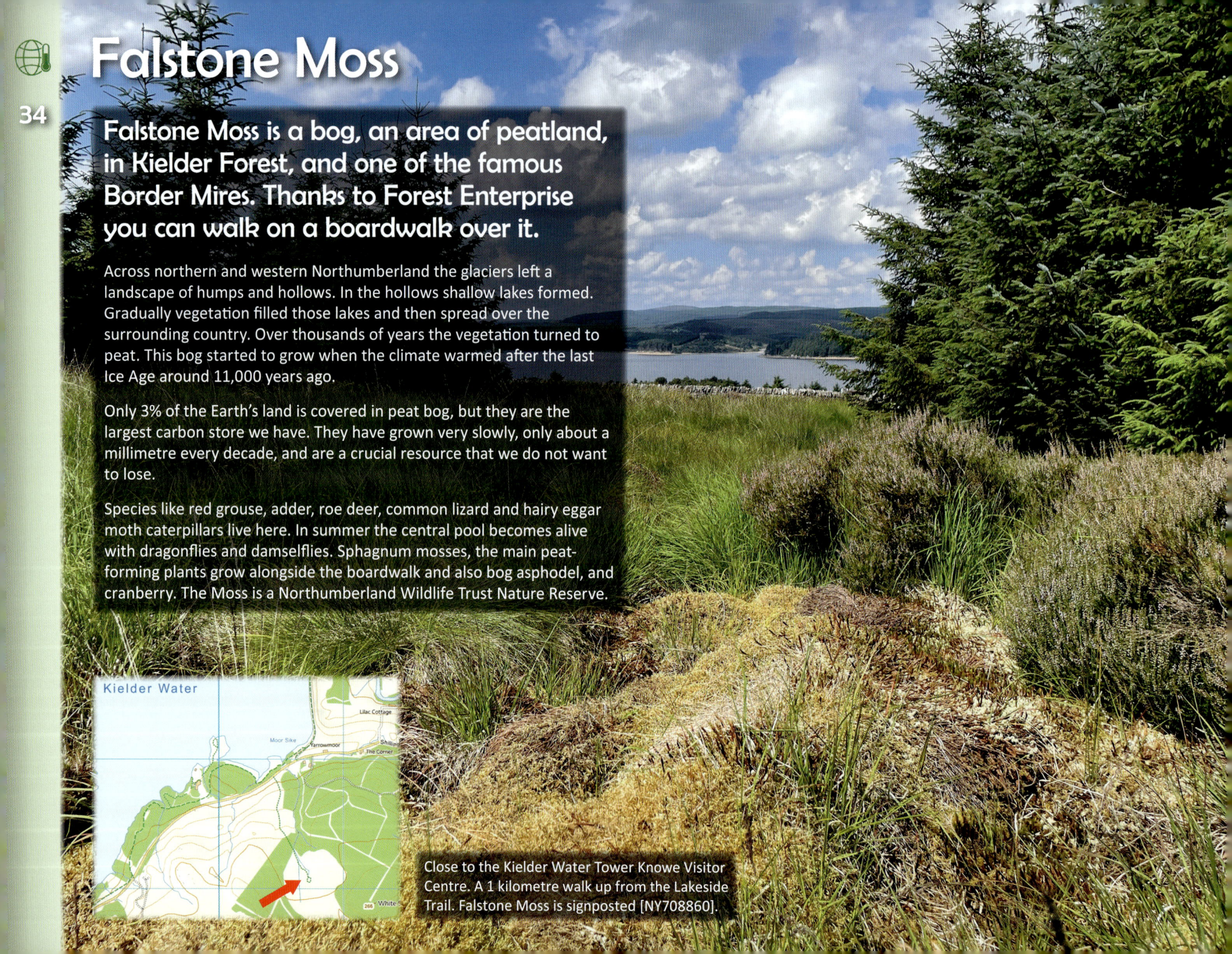

Falstone Moss is a bog, an area of peatland, in Kielder Forest, and one of the famous Border Mires. Thanks to Forest Enterprise you can walk on a boardwalk over it.

Across northern and western Northumberland the glaciers left a landscape of humps and hollows. In the hollows shallow lakes formed. Gradually vegetation filled those lakes and then spread over the surrounding country. Over thousands of years the vegetation turned to peat. This bog started to grow when the climate warmed after the last Ice Age around 11,000 years ago.

Only 3% of the Earth's land is covered in peat bog, but they are the largest carbon store we have. They have grown very slowly, only about a millimetre every decade, and are a crucial resource that we do not want to lose.

Species like red grouse, adder, roe deer, common lizard and hairy eggar moth caterpillars live here. In summer the central pool becomes alive with dragonflies and damselflies. Sphagnum mosses, the main peat-forming plants grow alongside the boardwalk and also bog asphodel, and cranberry. The Moss is a Northumberland Wildlife Trust Nature Reserve.

Close to the Kielder Water Tower Knowe Visitor Centre. A 1 kilometre walk up from the Lakeside Trail. Falstone Moss is signposted [NY708860].

Bog vegetation at Falstone Moss

Gelt Linn

High in the northern Pennines above Slaggyford is a fossil waterfall and steep valley, cut when there were melting ice sheets covering all northern England and huge torrents of water flowed under the ice and across the Pennine hills.

The last ice sheets were at their maximum around 25,000 years ago and then, over 10,000 years or so, started to melt away. The little stream, the Gelt Burn, that now flows in the valley and in a narrow slot gorge over the waterfall, is called a "misfit". That's because such a deep valley and big waterfall must have been cut by much more water than flows down them now.

The valley ("Butt Hill Channel") is here because it is one of the lowest points in between the Vale of Eden in the west and the South Tyne valley. The Vale of Eden was under an ice sheet more than a kilometre thick. Vast amounts of water under that ice needed to escape, so it flowed eastwards eroding a channel. We know the water was under the ice and under pressure because the channel sometimes runs uphill.

These are managed grouse moors so you will see grouse and in the spring and summer curlews, lapwings and golden plover. If you are lucky you may see short eared owls hunting across the heather.

The nearest place is Slaggyford, but it's quite a hike to the waterfall [NY639502] and even further to the channel NY632503], (about 12 kilometres and 16 kilometres round trips).

The fossil waterfall and slot gorge at Gelt Linn

Hauxley

In the small cliff below dune sands is layer of peat with ancient tree trunks sticking out of it. Below the peat is a stony clay laid down by an ice sheet during the last glaciation.

The coast is eroding here and new natural and human artifacts appear all the time, especially after winter storms. The peat and the tree trunks are roughly 7000 years old. The glacial clay (geologists used to call it boulder clay but now you will hear them talk about till or diamict) beneath was deposited during the last Ice Age which ended about 15,000 years ago. The sand dunes are more modern and still moving.

Archaeologists have also found human foot prints in the peat and mud on the foreshore. Beneath the sand dunes nearby they found many objects and relics left by peoples of many ages: from the Stone Age - Mesolithic period between 6,000 and 10,000 years ago, to Bronze Age, to medieval, and even bell pits used to mine coal beneath the rocky foreshore.

The peat was once vegetation in a low-lying poorly drained area. The trees, including oak, hazel and alder, are the remains of a forest. Sea level was lower then and the coastline was further east, in fact at one point in time you could have walked to the Netherlands! Stone Age people hunted in the forests and along the seashore.

The site is very close to Northumberland Wildlife Trust's nature reserve at Hauxley. The reserve was created after opencast coal extraction finished. It has a diverse flora and fauna of wetland, scrub, coastal and meadow species. Look out for the orange berries of the sea buckthorn and the birds, waterfowl and waders, which can be watched from the visitor centre and from hides by the ponds.

Hauxley Point, just east of Hauxley Nature Reserve, where you can park (NU285023).

Ancient tree trunk sticking out of a peat layer at Hauxley

Shaftoe Crags

In the middle of gentle, rolling Northumberland countryside, only 14 miles northwest of Newcastle are wild gritstone crags; a moorland oasis. The views from the crags are fantastic, with Simonside to the north, and Hadrian's Wall country to the south.

The gritstones of Shaftoe Crags are about 325 million years old and from the Carboniferous period. These and the crags to the north, near Rothley, are made of a coarse-grained, pebbly sandstone. They were once sand and pebbles carried in the channel of a large, fast-flowing river; over millions of years the sand and pebbles turned to rock. They are hard and so resist erosion and stand proud and high in the landscape. Some of the sandstone ridges are smoothed and orientated west-east, evidence of the scraping and streamlining by ice sheets.

Like many prominent rocky features, these crags have attracted the attention of humankind from Stone Age times and continue to do so. Look out for overhangs which may once have been rock shelters, the ancient Romano-British settlement at Salter's Nick and the drove road nearby. The Devil's Punch Bowl is a pothole on the crag top which folklore says was filled with wine to celebrate a Blackett wedding. Look for the Tailor and his Man, a huge split gritstone boulder at the base of the escarpment.

The rock faces are rich in ferns: all of brittle bladder-fern, maidenhair spleenwort, lady fern, male fern, black spleenwort, broad buckler-fern and common polypody can be found.

About 3 kilometres to the west of Bolam Lake Country Park [NZ053818], where you can park.

The Devil's Punch Bowl in the sandstone of Shaftoe Crags

Allenheads

Allenheads is an old lead mining and smelting village which sits at over 400 metres above sea level in the North Pennines. The lead workings are underground but, in the village and the surrounding countryside, evidence of lead mining and smelting is everywhere, from buildings, to capped old shafts, waste dumps and reservoirs.

It may seem a quiet place now but in the 18th and 19th centuries Allenheads was a busy and thriving town, producing more lead than anywhere else in the region. The veins which contain the lead (the ore mineral is called galena) originate from the injection of hot mineral-rich fluids 290 million years ago. Galena and other minerals, like fluorspar and quartz, originated deep underground. They cooled and crystallised out in cracks and crevices in the surrounding Carboniferous rocks, becoming veins in the limestones and sandstones.

The mine wastes contain lead, zinc and cadmium which are toxic to most plants. But spring sandwort, mountain pansy and Pyrenean scurvy-grass have a genetic tolerance to these elements, and are abundant on some of the former mining sites, although uncommon elsewhere.

Allenheads is high in the north Pennines. There is a car park in the centre of the village [NY860454], next to an information centre.

Old mining tubs in the centre of Allenheads

Beltingham river gravels

Just northwest of the small village of Beltingham is an area of river deposits – sand, pebbles and cobbles - beside the South Tyne. They are special because they contain the waste products of lead and zinc mining in the Pennine hills.

The plant colonies that are able to grow in these contaminated situations are very rare; they are called calaminarian habitats. Beltingham river gravels are a Northumberland Wildlife Trust Nature Reserve and Site of Special Scientific Interest.

Mining of lead and its associated minerals, silver, zinc and fluorite has taken place for many centuries in the hills and valleys of the northern Pennines. The lead and other heavy metals residues, like cadmium and barium, have been flushed far downstream. These metals, which are toxic to most plants, settle into the river deposits and the soils that develop on them.

The main heavy-metal tolerant plants are in quite a small area and are spring sandwort, alpine penny-cress, sea thrift (a variant), mountain pansy and dune helleborine. The condition of this environment is fragile and needs to be constantly monitored.

The river gravels are about 1 kilometre west and north of Beltingham village [NY785643], which is close to Bardon Mill.

The South Tyne at Beltingham

Duddo Stones

Today there are five standing stones up to 2.3 metres high, made of well weathered and naturally sculpted sandstone. There were originally 6 or 7 stones, then only 4 until a fifth was added in the early 20th century. The views over Northumberland are panoramic and dramatic, especially towards the Cheviot hills.

The stones were erected by Neolithic or Bronze Age people around 4000 years ago. The sandstone is Carboniferous in age and over 330 million years old. Early archaeologists found charcoal and pieces of bone in a pit in the middle of the stones. This might have been from a human burial. Modern experts say they do not fully understand exactly what these stone circles were for. But they must have been important ritual sites. Perhaps for the burial of the dead, or for calculating the seasons and astronomical events. Or perhaps they were a place that people would gather. Or maybe all of these reasons? What we do know is that, while it was for very different reasons, rocks and stones were just as important in their lives as they are in ours today.

The area directly around the stones has been left fallow, but the surrounding land is arable, growing cereal crops. Look out for bullfinches, redstarts and yellowhammers in the hedges. You may also see buzzards, kestrels, wheatears, linnets, tree sparrows, and meadow pipits.

A 2 kilometre walk north of the village of Duddo [NT930437].

Duddo Stones looking north

Dukesfield

The most striking remains of one of the region's largest lead smelting mills stands quietly in a wood beside the Devil's Water at Dukesfield.

Two elegant arches are the most visible evidence of an industry that was the backbone of the south Northumberland economy. It smelted lead ore (called galena), that was mined in the Pennine hills to the south, into ingots of lead.

The Dukesfield smelting mill was built around 1665 and continued to process lead and silver from ore until 1835. The veins of galena – the source of the lead and silver - were once hot, mineral-rich fluids, heated by a granite under the Pennines; the fluids filled cracks and fractures in Carboniferous sedimentary rocks around 290 million years ago.

The Dukesfield smelting mill was most active during the Napoleonic wars – between 1803 and 1815. This mill and others were in the ownership of the very old and very wealthy Blackett and Beaumont families.

It was sited where supplies of lead ore were near and the market and transport were also close – the wharves of Newcastle. Plus there was coal and water to power the machines and furnaces. Pack horses took the galena down from the hills to the smelter and then the refined lead onto Newcastle.

Despite consolidation works, tree seedlings of many different species are growing in the masonry as woodland tries to reclaim the site. A variety of herbs are also present – in late summer there are harebell and devil's-bit scabious, and heather nearby.

South of the hamlet called Juniper, which is south of Hexham. A 400 metre walk upstream from the bridge over the Devil's Water [NY942580].

The arches at Dukesfield, part of the former lead smelting mills

Hancock Museum

While the Hancock Museum has many wonderful things on display, from Roman artifacts, to life size models of modern animals, it is included in this book because its geology section houses some spectacular fossils, many of which were found in Northumberland.

The museum was named after two brothers, John and Albany Hancock. Albany was an expert on fossils, especially brachiopods, and on marine invertebrates, while John concentrated on birds. The Museum and its natural history and geology collections are owned by the Natural History Society of Northumbria, founded in 1829 and all but the oldest such society in the country.

Given that the most widespread rocks in Northumberland are Carboniferous, it is the fossils that are found in rocks of that age, 295 to 354 million years ago, that should perhaps hold most fascination for us Northumbrians. They range from huge trees and delicate plants that grew in ancient coal forests, to shells and corals that lived in primitive sub-tropical seas.

Arguably the most special fossil is the remains of one of the largest carnivorous amphibians which ever lived: Eogyrinus or Pholiderpeton. It exceeded 4.5 metres in length and hunted in the submerged and tangled roots of the swamps. Its bones were identified from Newsham colliery near Blyth by a Cramlington grocer and amateur palaeontologist called Thomas Atthey between 1850 and 1880.

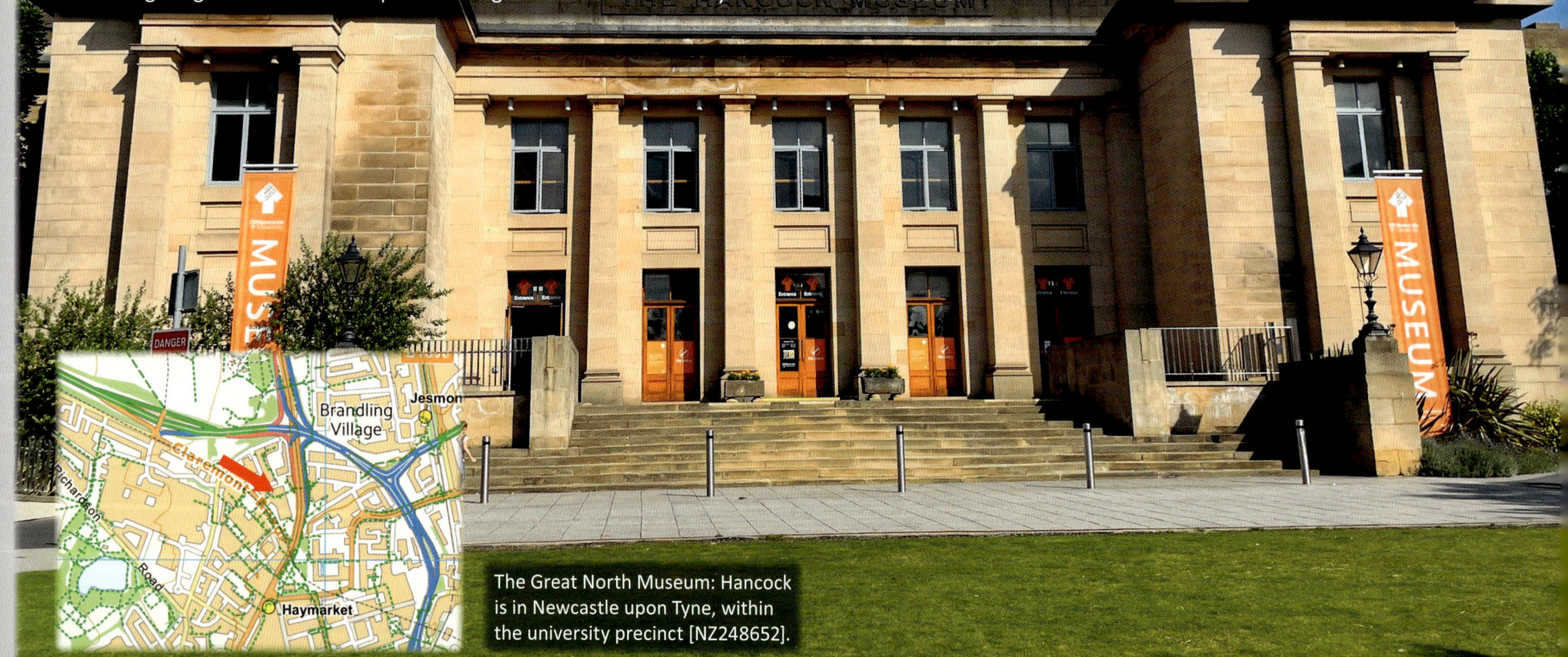

The Great North Museum: Hancock is in Newcastle upon Tyne, within the university precinct [NZ248652].

A model of a 320 million year old carnivorous amphibian (Eogyrinus)

Haltwhistle Burn

The trail up Haltwhistle Burn is a fantastic place to see the Carboniferous rocks that make up most of Northumberland and at the same time get a real insight into our industrial heritage. Every rock type here in the valley has been mined, or quarried.

The rocks date from between 320 and 330 million years ago. The gorge was largely cut when the last Ice Age finished around 15,000 years ago. When the ice melted it produced an enormous amount of meltwater. This cut a channel from the north to the River South Tyne. The rocks in the gorge – the sandstone, shales, limestones and coals – were once sands, muds, coral seas and swampy forests when Britain was only 5 degrees north of the Equator. At that time sea level rose and fell every 100,000 years or so, changing the environment from sandy river deltas, to swamps and forests, to muddy coastal lagoons and sub-tropical clear coral seas.

120 years ago there were active mines and quarries all along the burn, extracting coal, limestone for agriculture and mortar, sandstone for building and shale for bricks. Over 680 men and boys worked in the coal mine (South Tyne Colliery) at the south end of the burn. The shaft there was dug by hand and goes down more than 150m.

Some of the woodland is secondary after industry, but some is remnant ancient woodland. There are surviving ancient woodland species like the very attractive oak fern, as well as commoner ones like ramsons (wild garlic) and wood crane's-bill. Limestone outcrops have characteristic plants like fairy flax and mouse-ear hawkweed. You might see roe deer, squirrels, heron and dippers, and perhaps even an otter.

At the north end of Haltwhistle: Willia Road [NY708645]. Haltwhistle is served by both bus and rail services.

The Fell Chimney, East End Pit

Hareshaw Dene

Hareshaw Burn is a stream that cascades over many waterfalls in a beautiful wooded valley. There is a well signposted trail and at its northern end is the most famous of the waterfalls - Hareshaw Linn.

But at the southern end of the valley are clues to very different times in the 19th century, when this now peaceful place was the location for iron making on an industrial scale.

The rocks that the stream cuts through are Carboniferous in age; around 335 million years old. They are a series of harder sandstones and limestones and softer shales and siltstones, which erode at different rates and so form waterfalls.

The iron smelting industry exploited ironstone nodules that occurred in the shales in the valley and the other local natural resources of coal, limestone, sandstone, plus water for power. The smelting of iron took place for around a decade in 1840.

The iron working lasted for such a short time here because there was no easy way to transport the finished iron to markets, but also because of the financial problems of the owners. The rail connection to Bellingham arrived just after iron making had stopped, also by then higher quality ore and improved smelting equipment was available elsewhere.

The steep sides of the dene have preserved ancient woodland with special plants like herb Paris, toothwort, wood sanicle and goldilocks buttercup.

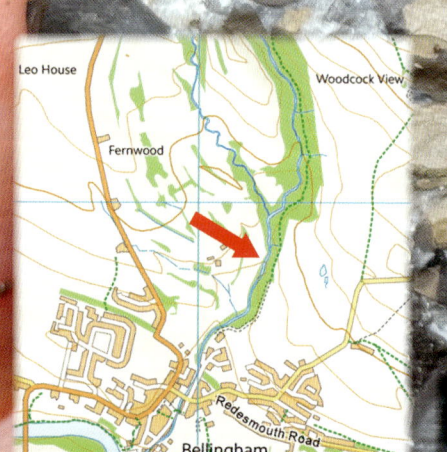

You can park in the Northumberland National Park Car Park [NY841835]. The walk to the Linn is about 5 kilometres there and back.

Hareshaw Linn: a waterfall over Carboniferous sandstones

Hartley New Pit

This small, tragic, memorial to so many lives lost is just one example of why it would be impossible to overstate the influence of coal and mining on the culture, economy and heritage of Northumberland. That influence extends from the coast, to the Tyne Valley, to the Pennines.

One of the worst accidents in the exploitation of coal in Northumberland took place at New Hartley. The disaster took place on 16 January 1862. 204 men and boys died when the beam of the pit's pumping engine broke and fell down the single mine shaft, blocking it and trapping the miners. It is hard to remain unmoved by the long list of names of the dead, their ages, and their family ties.

A second memorial was opened on the site of Hartley New Pit (also known as Hester Pit) in 1976. The dead are buried in Earsdon churchyard where the obelisk records their names and ages. The coal seams that Hartley New Pit mined are, like the rest of the Northumberland and Durham coalfield, Carboniferous and are around 320 million years old. After the disaster the pit was closed.

There was one very significant positive outcome of the Hartley disaster. No colliery would in future be allowed to have only a single shaft; all collieries would by law have to have two shafts.

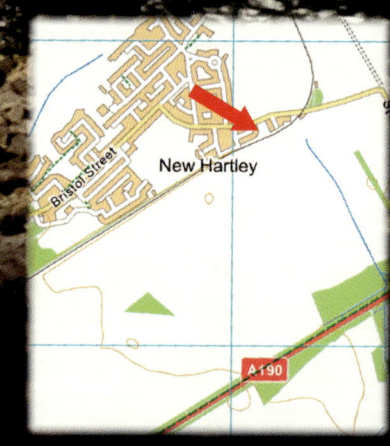

The New Hartley memorial is just northeast of Seaton Delaval [NZ312767]. The obelisk is at Earsdon [NZ321726].

ERECTED

TO THE MEMORY OF THE 204 MINERS, WHO LOST THEIR LIVES IN HARTLEY PIT, BY THE FATAL CATASTROPHE OF THE ENGINE BEAM BREAKING, 16TH JANUARY 1862.

Jᵒ AMOUR	AGED 43	Cᵉ WANLESS	AGED 20
Rᵈ AMOUR	AGED 14	Tᵒ WANLESS	AGED 19
Jᵉ TERNENT	AGED 44	Jⁿ WANLESS	AGED 14
Gᵉ TERNENT	AGED 15	Wᵐ JACK	AGED 24
Wᵐ PAPE	AGED 14	Wᵐ GLEDSON	AGED 71
Tᵒ SHARP	AGED 48	Wᵐ GLEDSON	AGED 43
Hʸ SHARP	AGED 44	Gᵉ GLEDSON	AGED 41
Aⁿ ELLIOTT	AGED 29	Tᵒ GLEDSON	AGED 36
Gᵉ SHARP	AGED 49	Tᵒ GLEDSON	AGED 16
Gᵉ SHARP	AGED 15	Wᵐ LIDDLE	AGED 40
Jⁿ SHARP	AGED 13	Wᵐ LIDDLE	AGED 17
Jᵒ BEWICK	AGED 34	Jᵒ LIDDLE	AGED 15
Jᵒ BEWICK	AGED 32	Jⁿ LIDDLE	AGED 10
Rᵒ BEWICK	AGED 30	Tᵒ LIDDLE	AGED 18
Tᵒ ROBINSON	AGED 42	Gᵉ LIDDLE	AGED 16
Tᵒ DAWSON	AGED 49	Jᵒ LIDDLE	AGED 11
Jⁿ DAWSON	AGED 12	Tᵒ LIDDLE	AGED 11
Aᵇ RICHARDSON	AGED 22	Tᵒ LIDDLE	AGED 18
Jᵒ JOHNSON	AGED 41	Tᵒ LAWS	AGED
Rᵗ JOHNSON	AGED 42	Gᵉ LAWS	AGED 23
Tᵒ COAL	AGED 37	Wᵐ LOUGE	AGED 30
Tᵒ CHAMBERS	AGED 55	Jᵒ LONG	AGED 15
Gᵉ CHAMBERS	AGED 19	Rᵒ LONG	AGED 17
Jᵒ HUMBLE	AGED 27	M. MURRAY	AGED 28
Wᵐ DIXON	AGED 34	R. MURLEY	AGED 23

The memorial obelisk in Earsdon churchyard

Lynemouth Beach

Along many beaches on the Northumberland coast, from Tynemouth to Berwick-upon-Tweed, are stretches of coal and black coaly shale debris in the sand. At Lynemouth Beach they can be extensive.

The coal and shale are Carboniferous – so around 320 million years old, but they have been washed up by recent tide and wave action. The coal comes mainly from colliery waste which only a few decades ago was dumped into the sea from coal mines near the coast. Coal is lighter than almost all other rocks and so it separates out and gets deposited last by the tide. The other source of some of the coal could be coastal and undersea outcrops of coal seams.

Sea coal may have a wider meaning elsewhere and historically, but here in the northeast it describes the coal deposited on the beaches, which was shovelled into horse-drawn carts taken down to the edge of the sea.

In the summer of 2021 man-made waste other than coal was still evident here; an old colliery landfill site was actively being eroded by the sea. So while this may not be the prettiest photograph in the book, it does serve to remind us of the enduring damage we are capable of causing to this precious environment, and that humans are, knowingly and unknowingly, creating their own unappealing geological period – the Anthropocene. A clean-up operation is being undertaken by Northumberland County Council, but until it is complete this is not the most picturesque place to watch rocks or wildlife.

Lynemouth Beach is about 5 kilometres northeast of Ashington [NZ305910].

Sea coal on the beach at Lynemouth

Newcastle upon Tyne

The stones used in the buildings of every town give us a chance to see an amazing variety of rocks and we don't have to travel very far. The city of Newcastle upon Tyne is no exception.

The stones vary from sedimentary, to metamorphic, to igneous and from very old to geologically young. The sources of stone are equally diverse, from quarries as near as Kenton, to as far away as Norway.

The building stones of the Civic Centre have a huge range; from the 460 million year old metamorphic rocks (schists from Otta in Norway) behind the River God Tyne, to the 150 million year old white, fossiliferous Jurassic limestones, from Portland on the south coast, which face the tower.

Granite features often; the plinth of the fantastic 1914 war memorial, The Response, at Barras Bridge is granite from Shap in Cumbria, as are the entrances to Metro stations. But the most famous streets and buildings in Newcastle, and the most architecturally outstanding, Grey Street and the rest of Grainger Town, are all constructed of local Carboniferous sandstones from places, like Denwick, Heddon, Springwell, Fourstones and Kenton.

Stone has long been the preferred choice for building in the north of Britain, where there are many local sources of suitable rocks. 1900 years ago Hadrian built his Wall of local sandstone. The 19th century architects and developers, like Dobson and Grainger, chose local sandstone too for their neo-classical rebuilding of the city centre.

The centre of the city of Newcastle upon Tyne [NZ248644].

The River God Tyne and Otta schist slabs facing the building

The Spetchells

Squeezed between the River Tyne and the Newcastle – Carlisle rail line at Prudhoe is a wooded mound, a ridge really, almost 1.5 kilometres long, 100 metres wide and 20 metres high. Where the vegetation and soil have been worn away, it is white.

The ridge, known as The Spetchells (the previous name of the local riverside), is man-made. It is made of millions of tons of waste material from a factory in Prudhoe which produced ammonium sulphate for explosives and fertilizer during World War 2. It was turfed to disguise the ridge and the factory from German bombers (the trees came later). The ridge is made of calcium carbonate, the same composition as limestone and chalk and so this is the only piece of "Chalk downland" in the county.

The Spetchells is a special habitat here in Northumberland. It hasn't exactly got a southern chalk grassland flora but rather a mix of lime-loving plants. They include musk-mallow, must thistle, kidney vetch, wild marjoram and traveller's-joy.

You can take the train to Prudhoe Station, or park at Tyne Riverside Country Park beside the bridge to Ovingham and take the short walk east to The Spetchells. [NZ097641].

On top of the Spetchells ridge

Town Moor

In the middle of Newcastle is a huge area of pasture land, much larger than Central Park in New York. What gets the Town Moor and its neighbours, Nuns Moor and Hunters Moor, into this rock book? They are one of the very few places where you can see evidence of the mining of coal in the city whose name is synonymous the world over with coal.

The coal seam mined directly underneath the moor is Carboniferous and around 312 million years old. It is called the High Main seam and is a very thick coal – sometimes exceeding 2 metres. That thickness and the fact that it was so close to the surface made it very attractive to men wishing to mine coal. The mining here may have started as early as the 14th century. In the 18th and 19th centuries mining was extensive by bell pits and then pillar and stall methods. The very uneven surface of Nuns Moor west of Grandstand Road is caused by old coal workings and their collapse - subsidence. As recently as the Second World War, because of the urgent need for fuel and power, over 340,000 square metres of coal was mined by opencast methods in the northern parts of the moors.

The Moor is mainly agricultural grassland, supporting cattle (within mooing distance of St James' Park), and there is also some woodland. Open country birds include skylark, and in winter waders such as golden plover. In the woodland are willow warbler, chiffchaff and blackcap.

A 2.5 kilometre walk north of Newcastle city centre [NZ233663].

Road A167

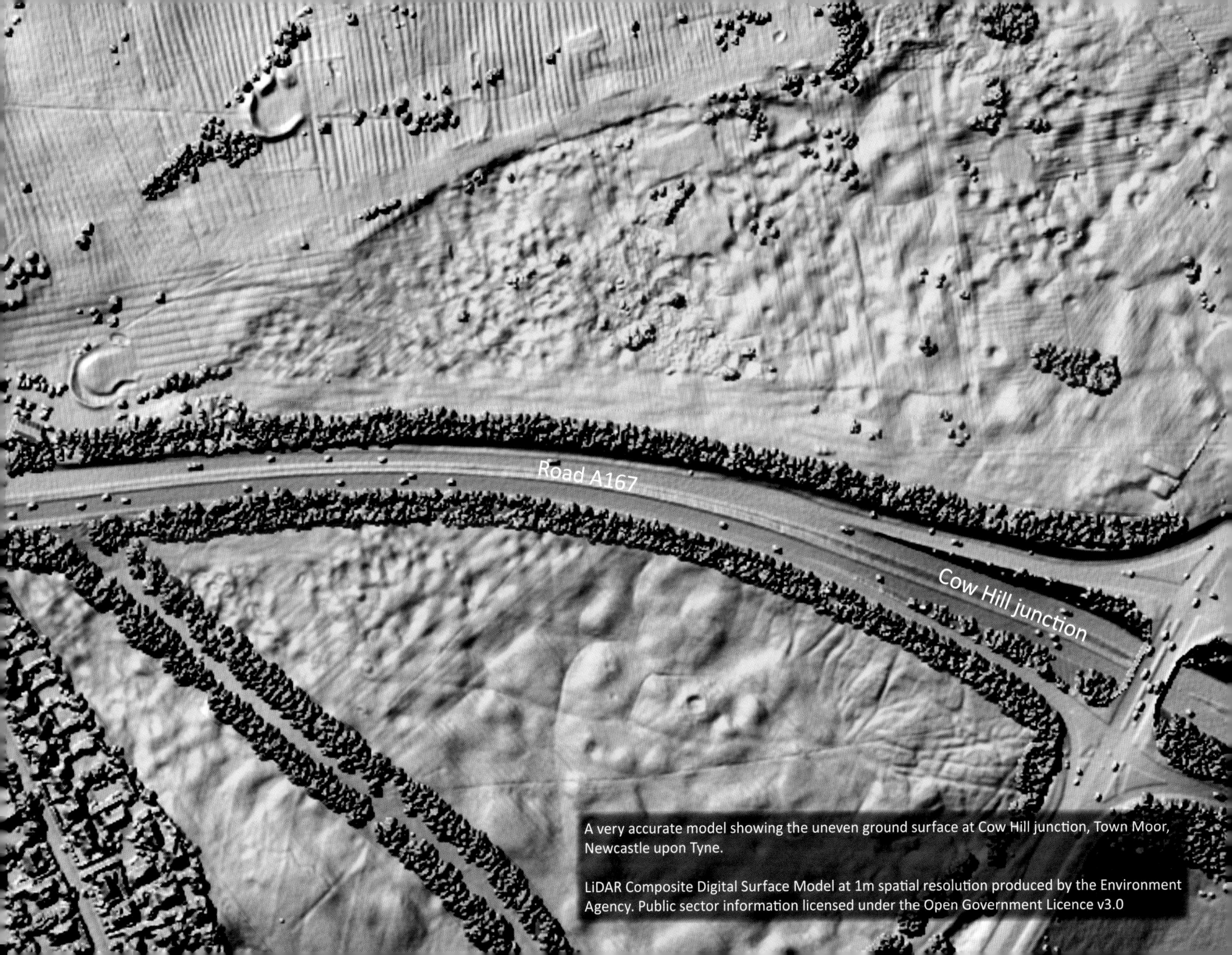

Cow Hill junction

A very accurate model showing the uneven ground surface at Cow Hill junction, Town Moor, Newcastle upon Tyne.

LiDAR Composite Digital Surface Model at 1m spatial resolution produced by the Environment Agency. Public sector information licensed under the Open Government Licence v3.0

Woodhorn Museum

Woodhorn Museum was once a coal mine. It now tells the story of what it was like to be a coal miner and also paints a picture of the lives of their families and the decline of the mining industry in Northumberland.

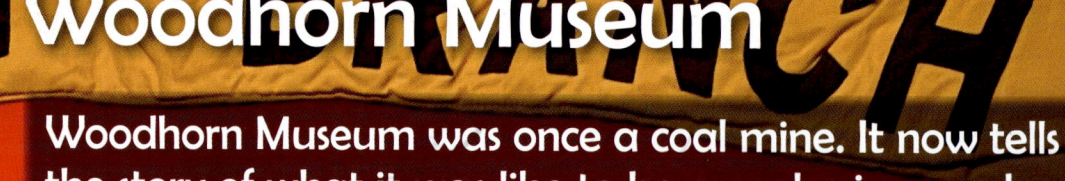

The coal seams that Woodhorn Colliery mined are Carboniferous and around 312 million years old. The mine itself began in 1894 and closed in 1981 as workable resources of coal declined. It is one of the few coal mine sites where the pit headgear and winding houses still survive – and can be visited. Woodhorn Colliery was one of over 600 collieries in the Northumberland and Durham Coalfield. There are none today. The last one to close in Northumberland was Ellington in 2005. While coal and other fossil fuels are now, rightly, regarded as the cause of our changing climate, their impact on the culture and heritage of Northumberland should never be underestimated.

Like all other pits Woodhorn was an area of heavy and dirty industry. It had its own pit heap where the waste from mining was dumped. While the pits were working these were barren areas, devoid of much, if any, vegetation and wildlife. Since closure Woodhorn colliery has been intensively landscaped, with planting of amenity woodland and a manicured meadow. Common birds such as long-tailed tit, jay and sparrowhawk may be found as well as a few woodland plants along the margins. The meadows, and other open areas, such as the narrow gauge railway track have interesting plants such as yellow-wort, and attract many bees and hoverflies. The adjacent lake at QEII Country Park was once the spoil heap but now supports a variety of water birds including great crested grebe. Its margins host emergent plants such as water mint and branched bur-reed, as well as damselflies and dragonflies.

Just northeast of Ashington on the A189 [NZ289885].

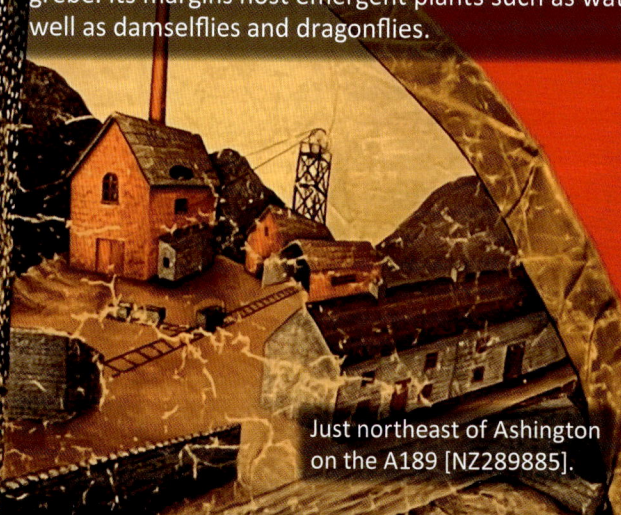

The twin coal mine shafts at Woodhorn Museum

Endnote

The aim of this book is to encourage you to look at your landscape through different eyes. To glimpse beneath the surface and reflect on its evolution and the ancient origins of its terrains and its habitats – our biodiversity depends on geodiversity. Equally important, I hope that the book also shows how changing climates and environments throughout geological time have shaped this precious landscape..... and are still changing it.

I hope it inspires you to look afresh at places you may know well, but also to explore less well-known places. Places where earthquakes have bent solid rock, where you can climb across the remnants of 400 million year-old volcanos, search for fossils and agates, or walk over a Border Mire that started when the last glacier left 15,000 years ago. Or simply wonder at the ingenuity of our ancestors who built enduring cairns, walls, roads and castles, and mined the stones that have created our shared heritage.

Alwin valley